Springer Tracts in Modern Physics
Volume 173

Managing Editor: G. Höhler, Karlsruhe

Editors: J. Kühn, Karlsruhe
Th. Müller, Karlsruhe
A. Ruckenstein, New Jersey
F. Steiner, Ulm
J. Trümper, Garching
P. Wölfle, Karlsruhe

Honorary Editor: E. A. Niekisch, Jülich

Now also Available Online

Starting with Volume 165, Springer Tracts in Modern Physics is part of the Springer LINK service. For all customers with standing orders for Springer Tracts in Modern Physics we offer the full text in electronic form via LINK free of charge. Please contact your librarian who can receive a password for free access to the full articles by registration at:

http://link.springer.de/series/stmp/reg_form.htm

If you do not have a standing order you can nevertheless browse through the table of contents of the volumes and the abstracts of each article at:

http://link.springer.de/series/stmp/

There you will also find more information about the series.

Springer
Berlin
Heidelberg
New York
Barcelona
Hong Kong
London
Milan
Paris
Singapore
Tokyo

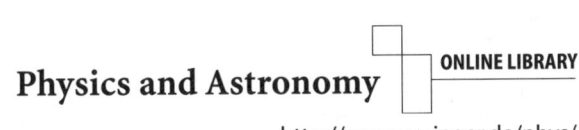

Springer Tracts in Modern Physics

Springer Tracts in Modern Physics provides comprehensive and critical reviews of topics of current interest in physics. The following fields are emphasized: elementary particle physics, solid-state physics, complex systems, and fundamental astrophysics.
Suitable reviews of other fields can also be accepted. The editors encourage prospective authors to correspond with them in advance of submitting an article. For reviews of topics belonging to the above mentioned fields, they should address the responsible editor, otherwise the managing editor. See also http://www.springer.de/phys/books/stmp.html

Managing Editor

Gerhard Höhler

Institut für Theoretische Teilchenphysik
Universität Karlsruhe
Postfach 69 80
76128 Karlsruhe, Germany
Phone: +49 (7 21) 6 08 33 75
Fax: +49 (7 21) 37 07 26
Email: gerhard.hoehler@physik.uni-karlsruhe.de
http://www-ttp.physik.uni-karlsruhe.de/

Elementary Particle Physics, Editors

Johann H. Kühn

Institut für Theoretische Teilchenphysik
Universität Karlsruhe
Postfach 69 80
76128 Karlsruhe, Germany
Phone: +49 (7 21) 6 08 33 72
Fax: +49 (7 21) 37 07 26
Email: johann.kuehn@physik.uni-karlsruhe.de
http://www-ttp.physik.uni-karlsruhe.de/~jk

Thomas Müller

Institut für Experimentelle Kernphysik
Fakultät für Physik
Universität Karlsruhe
Postfach 69 80
76128 Karlsruhe, Germany
Phone: +49 (7 21) 6 08 35 24
Fax: +49 (7 21) 6 07 26 21
Email: thomas.muller@physik.uni-karlsruhe.de
http://www-ekp.physik.uni-karlsruhe.de

Fundamental Astrophysics, Editor

Joachim Trümper

Max-Planck-Institut für Extraterrestrische Physik
Postfach 16 03
85740 Garching, Germany
Phone: +49 (89) 32 99 35 59
Fax: +49 (89) 32 99 35 69
Email: jtrumper@mpe-garching.mpg.de
http://www.mpe-garching.mpg.de/index.html

Solid-State Physics, Editors

Andrei Ruckenstein
Editor for The Americas

Department of Physics and Astronomy
Rutgers, The State University of New Jersey
136 Frelinghuysen Road
Piscataway, NJ 08854-8019, USA
Phone: +1 (732) 445 43 29
Fax: +1 (732) 445-43 43
Email: andreir@physics.rutgers.edu
http://www.physics.rutgers.edu/people/pips/
Ruckenstein.html

Peter Wölfle

Institut für Theorie der Kondensierten Materie
Universität Karlsruhe
Postfach 69 80
76128 Karlsruhe, Germany
Phone: +49 (7 21) 6 08 35 90
Fax: +49 (7 21) 69 81 50
Email: woelfle@tkm.physik.uni-karlsruhe.de
http://www-tkm.physik.uni-karlsruhe.de

Complex Systems, Editor

Frank Steiner

Abteilung Theoretische Physik
Universität Ulm
Albert-Einstein-Allee 11
89069 Ulm, Germany
Phone: +49 (7 31) 5 02 29 10
Fax: +49 (7 31) 5 02 29 24
Email: steiner@physik.uni-ulm.de
http://www.physik.uni-ulm.de/theo/theophys.html

G. Alber T. Beth M. Horodecki P. Horodecki
R. Horodecki M. Rötteler
H. Weinfurter R. Werner A. Zeilinger

Quantum Information

An Introduction to Basic Theoretical Concepts and Experiments

With 60 Figures

 Springer

The corresponding author:

Dr. Gernot Alber
Universität Ulm
Abteilung für Quantenphysik
Albert-Einstein-Allee 11
89069 Ulm, GERMANY
E-mail: gernot.alber@physik.uni-ulm.de

The complete list of authors see page XI

Library of Congress Cataloging-in-Publication Data.
Quantum information: an introduction to basic theoretical concepts and experiments/G. Alber ... [et al.]. p.cm. – (Springer tracts in modern physics; 173). Includes bibliographical references and index. ISBN 3-540-41666-8 (alk. paper) 1. Quantum computers. I. Alber, Gernot. II. Series. QA76.889.Q848 2001   004.1–dc21   2001020642

Physics and Astronomy Classification Scheme (PACS): 03.67.-a, 03.67.Hk, 03.67.Lx, 03.67.Dd

ISSN print edition: 0081-3869
ISSN electronic edition: 1615-0430
ISBN 3-540-41666-8 Springer-Verlag Berlin Heidelberg New York

This work is subject to copyright. All rights are reserved, whether the whole or part of the material is concerned, specifically the rights of translation, reprinting, reuse of illustrations, recitation, broadcasting, reproduction on microfilm or in any other way, and storage in data banks. Duplication of this publication or parts thereof is permitted only under the provisions of the German Copyright Law of September 9, 1965, in its current version, and permission for use must always be obtained from Springer-Verlag. Violations are liable for prosecution under the German Copyright Law.

Springer-Verlag Berlin Heidelberg New York
a member of BertelsmannSpringer Science+Business Media GmbH

http://www.springer.de

© Springer-Verlag Berlin Heidelberg 2001
Printed in Germany

The use of general descriptive names, registered names, trademarks, etc. in this publication does not imply, even in the absence of a specific statement, that such names are exempt from the relevant protective laws and regulations and therefore free for general use.

Typesetting: Camera-ready copy from the authors using a Springer LaTeX macro package
Cover design: *design & production* GmbH, Heidelberg

Printed on acid-free paper   SPIN: 10791899   56/3141/tr   5 4 3 2 1 0

Preface

Though quantum theory is celebrating its 100th anniversary this year, quantum information processing is still a remarkably young research field. The questions driving this research field reflect a profound change in the general attitude towards the fundamental aspects of quantum theory. So far, research on the foundations of quantum theory has been concerned mainly with the theoretical exploration of those particular features which distinguish quantum theory from classical physics. The main intention of quantum information processing is to exploit these specific features for technological purposes. As early as 1935, Erwin Schrödinger had already noted that one of these characteristic features of quantum theory is the phenomenon of entanglement. Many years passed from this early insight until John Bell realized the quantitative consequences of the corresponding quantum correlations in his famous work from 1964. These theoretical predictions inspired numerous experiments, which all support the peculiar features predicted for quantum correlations. From these purely theoretical insights, it again required a long period of development to arrive at those potentially useful applications which are now of central interest for the processing of quantum information.

The following contributions provide an introductory overview of basic problems, methods and topical results in this research field. The idea of producing this volume was born at a symposium on this subject which was held at the 1999 annual spring meeting of the Deutsche Physikalische Gesellschaft in Heidelberg. This symposium was organized jointly by the Quantum Optics and Mathematical Physics sections. The widespread interest, the success of this symposium and the initiative of Prof. Frank Steiner, the head of the Mathematical Physics section, motivated us to edit a volume on basic problems, methods and recent results in this rapidly evolving field. This book should be useful for students and active researchers in physics, computer science and mathematics who want to learn about the most recent developments in this exciting research field.

Ulm, March 2001 *Gernot Alber*

Contents

**1 From the Foundations of Quantum Theory
to Quantum Technology – an Introduction**
Gernot Alber 1

1.1 Early Developments 2
    1.1.1 Entanglement and Local, Realistic Theories 2
    1.1.2 Characteristic Quantum Effects for Practical Purposes 6
    1.1.3 Quantum Algorithms 8
1.2 Quantum Physics and Information Processing 11

2 Quantum Information Theory – an Invitation
Reinhard F. Werner 14

2.1 Introduction 14
2.2 What is Quantum Information? 15
2.3 Impossible Machines 17
    2.3.1 The Quantum Copier 18
    2.3.2 Joint Measurement 19
    2.3.3 Bell's Telephone 19
    2.3.4 Entanglement, Mixed-State Analyzers
    and Correlation Resolvers 23
2.4 Possible Machines 25
    2.4.1 Operations on Multiple Inputs 25
    2.4.2 Quantum Cryptography 27
    2.4.3 Entanglement-Assisted Teleportation 27
    2.4.4 Superdense Coding 29
    2.4.5 Quantum Computation 29
    2.4.6 Error Correction 30
2.5 A Preview of the Quantum Theory of Information 31
2.6 Elements of Quantum Information Theory 34
    2.6.1 Systems and States 34
    2.6.2 Channels 39
    2.6.3 Duality between Channels and Bipartite States 45
    2.6.4 Channel Capacity 47
        Comparison with the Classical Definition 50
        Comparison with other Error Criteria 51

2.6.5 Coding Theorems	51
2.6.6 Teleportation and Dense Coding Schemes	53

3 Quantum Communication
Harald Weinfurter and Anton Zeilinger 58

3.1 Introduction	58
3.2 Entanglement – Basic Features	60
3.3 The Quantum Communication Schemes	62
3.3.1 Quantum Cryptography	63
3.3.2 Quantum Dense Coding	66
3.3.3 Quantum Teleportation	68
The Idea	68
Some Remarks	70
3.4 The Experimental Prerequisites	73
3.4.1 Entangled Photon Pairs	74
3.4.2 Polarization-Entangled Pairs from Type II Down-Conversion	76
3.4.3 Interferometric Bell-State Analysis	79
The Principle	79
Bell-State Analysis of Independent Photons	81
3.4.4 Manipulation and Detection of Single Photons	82
3.5 Quantum Communication Experiments	83
3.5.1 Quantum Cryptography	83
3.5.2 Quantum Dense Coding	85
3.5.3 Quantum Teleportation of Arbitrary Qubit States	89
3.6 Outlook	93

4 Quantum Algorithms: Applicable Algebra and Quantum Physics
Thomas Beth and Martin Rötteler 96

4.1 Introduction	96
4.2 Architectures and Machine Models	97
4.2.1 Quantum Networks	98
4.2.2 Boolean Functions and the Ring Normal Form	103
4.2.3 Embedded Transforms	104
4.2.4 Permutations	106
4.2.5 Preparing Quantum States	108
4.2.6 Quantum Turing Machines	109
4.3 Using Entanglement for Computation: A First Quantum Algorithm	113
4.4 Quantum Fourier Transforms: the Abelian Case	115
4.4.1 Factorization of DFT_N	116
4.4.2 Abelian Groups and Duality Theorems	117
The Dual Group	118
4.4.3 Sampling of Fourier Coefficients	119

        4.4.4 Schur's Lemma and its Applications
              in Quantum Computing 120
   4.5 Exploring Quantum Algorithms 122
        4.5.1 Grover's Algorithm 122
        4.5.2 Shor's Algorithm 124
              Diophantine Approximation 127
        4.5.3 Taxonomy of Quantum Algorithms........................... 128
              Entanglement-Driven Algorithms 128
              Superposition-Driven Algorithms 129
   4.6 Quantum Signal Transforms 129
        4.6.1 Quantum Fourier Transforms: the General Case 129
              The Wedderburn Decomposition 130
              A Decomposition Algorithm................................ 131
              An Example: Wreath Products 133
              Nonabelian Hidden Subgroups.............................. 135
        4.6.2 The Discrete Cosine Transform 137
   4.7 Quantum Error-Correcting Codes 140
        4.7.1 Introduction 140
        4.7.2 Background 141
        4.7.3 A Classic Code 142
        4.7.4 Quantum Channels and Codes 146
              Quantum Channels 146
              Quantum Codes 147
   4.8 Conclusions.. 150

5 **Mixed-State Entanglement and Quantum Communication**
   Michał Horodecki, Paweł Horodecki and Ryszard Horodecki 151

   5.1 Introduction 151
   5.2 Entanglement of Mixed States: Characterization................. 153
        5.2.1 Pure States 154
        5.2.2 Some Necessary Conditions for Separability
              of Mixed States... 155
        5.2.3 Entanglement and Theory of Positive Maps 157
              Positive and Completely Positive Maps 157
              Characterization of Separable States **via** Positive Maps 159
              Operational Characterization of Entanglement
                  in Low Dimensions ($2 \otimes 2$ and $2 \otimes 3$ Systems) 160
              Higher Dimensions – Entangled States
                  with Positive Partial Transposition 161
              Range Criterion and Positive Nondecomposable Maps 163
        5.2.4 Examples 164
              Reduction Criterion for Separability..................... 165
              Strong Separability Criteria from an Entanglement Witness .. 165
              Werner States 165

X   Contents

    Isotropic State 166
    A Two-Qubit State 167
    Entangled PPT State Via Nondecomposable Positive Map ... 167
  5.2.5 Volumes of Entangled and Separable States 168
5.3 Mixed-State Entanglement as a Resource
  for Quantum Communication 170
  5.3.1 Distillation of Entanglement:
    Counterfactual Error Correction 170
  5.3.2 Distillation of Two-Qubit States 173
    BBPSSW Distillation Protocol 173
    All Entangled Two-Qubit States are Distillable 175
  5.3.3 Examples 177
    Distillation of Isotropic State for $d \otimes d$ System 177
    Distillation and Reduction Criterion 177
  5.3.4 Bound Entanglement 178
  5.3.5 Do There Exist Bound Entangled NPT States? 181
  5.3.6 Example 183
  5.3.7 Some Consequences of the Existence
    of Bound Entanglement 183
    Bound Entanglement and Teleportation 183
    General Teleportation Scheme 184
    Optimal Teleportation 184
    Teleportation Via Bound Entangled States 185
    Activation of Bound Entanglement 185
    Conclusive Teleportation 185
    Activation Protocol 187
    Entanglement-Enhanced LOCC Operations 190
    Bounds for Entanglement of Distillation 191
5.4 Concluding Remarks 193

References 197

Index 211

List of Authors

Gernot Alber
Abteilung für Quantenphysik
Universität Ulm
Albert-Einstein-Allee 11
89069 Ulm, Germany
gernot.alber@physik.uni-ulm.de

Thomas Beth
Institut für Algorithmen
und Kognitive Systeme
Universität Karlsruhe
Am Fasanengarten 5
76128 Karlsruhe, Germany
EISS_Office@ira.uka.de

Michał Horodecki
Institute of Theoretical Physics
and Astrophysics
University of Gdansk
ul. Wita Stwosza 51
80–952 Gdansk, Poland
michalh@iftia.univ.gda.pl

Paweł Horodecki
Dept. of Technical Physics
and Applied Mathematics
Technical University of Gdansk
ul. Narutowicza 11/12
81–952 Gdansk, Poland
pawel@mifgate.pg.gda.pl

Ryszard Horodecki
Institute of Theoretical Physics
and Astrophysics
University of Gdansk
ul. Wita Stwosza 51
80–952 Gdansk, Poland
fizrh@univ.gda.pl

Martin Rötteler
Institut für Algorithmen
und Kognitive Systeme
Universität Karlsruhe
Am Fasanengarten 5
762128 Karlsruhe, Germany
martin.roetteler
@informatik.uni-karlsruhe.de

Harald Weinfurter
Sektion Physik
Universität München
Schellingstr. 4
80797 München, Germany
harald.weinfurter
@physik.uni-muenchen.de

Reinhard Werner
Institut für Mathematische Physik
TU Braunschweig
Mendelsohnstr. 3
38106 Braunschweig, Germany
r.werner@tu-bs.de

Anton Zeilinger
Institut für Experimentalphysik
Universität Wien
Boltzmanngasse 5
1090 Wien, Austria
anton.zeilinger@univie.ac.at

1 From the Foundations of Quantum Theory to Quantum Technology – an Introduction

Gernot Alber

Nowadays, the new technological prospects of processing quantum information in quantum cryptography [1], quantum computation [2] and quantum communication [3] attract not only physicists but also researchers from other scientific communities, mainly computer scientists, discrete mathematicians and electrical engineers. Current developments demonstrate that characteristic quantum phenomena which appear to be surprising from the point of view of classical physics may enable one to perform tasks of practical interest better than by any other known method. In quantum cryptography, the no-cloning property of quantum states [4] or the phenomenon of entanglement [5] helps in the exchange of secret keys between various parties, thus ensuring the security of one-time-pad cryptosystems [6]. Quantum parallelism [7], which relies on quantum interference and which typically also involves entanglement [8], may be exploited for accelerating computations. Quantum algorithms are even capable of factorizing numbers more efficiently than any known classical method is [9], thus challenging the security of public-key cryptosystems such as the RSA system [6]. Classical information and quantum information based on entangled quantum systems can be used for quantum communication purposes such as teleporting quantum states [10, 11].

Owing to significant experimental advances, methods for processing quantum information have developed rapidly during the last few years.[1] Basic quantum communication schemes have been realized with photons [10,11], and basic quantum logical operations have been demonstrated with trapped ions [13,14] and with nuclear spins of organic molecules [15]. Also, cavity quantum electrodynamical setups [16], atom chips [17], ultracold atoms in optical lattices [18,19], ions in an array of microtraps [20] and solid-state devices [21–23] are promising physical systems for future developments in this research area. All these technologically oriented, current developments rely on fundamental quantum phenomena, such as quantum interference, the measurement process and entanglement. These phenomena and their distinctive differences from basic concepts of classical physics have always been of central interest in research on the foundations of quantum theory. However, in emphasizing their technological potential, the advances in quantum infor-

---

[1] Numerous recent experimental and theoretical achievements are discussed in [12].

mation processing reflect a profound change in the general attitude towards these fundamental phenomena. Thus, after almost two decades of impressive scientific achievements, it is time to retrace some of those significant early developments in quantum physics which are at the heart of quantum technology and which have shaped its present-day appearance.

1.1 Early Developments

Many of the current methods and developments in the processing of quantum information have grown out of a long struggle of physicists with the foundations of modern quantum theory. The famous considerations by Einstein, Podolsky and Rosen (EPR) [24] on reality, locality and completeness of physical theories are an early example in this respect. The critical questions raised by these authors inspired many researchers to study quantitatively the essential difference between quantum physics and the classical concepts of reality and locality. The breakthrough was the discovery by J.S. Bell [25] that the statistical correlations of entangled quantum states are incompatible with the predictions of any theory which is based on the concepts of reality and locality of EPR. The constraints imposed on statistical correlations within the framework of a local, realistic theory (LRT) are expressed by Bell's inequality [25]. As the concept of entanglement and its peculiar correlation properties have been of fundamental significance for the development of quantum information processing, it is worth recalling some of its most elementary features in more detail.

1.1.1 Entanglement and Local, Realistic Theories

In order to clarify the characteristic differences between quantum mechanical correlations originating from entangled states and classical correlations originating from local, realistic theories, let us consider the following basic experimental setup (Fig. 1.1). A quantum mechanical two-particle system, such as a photon pair, is produced by a source s. Polarization properties of

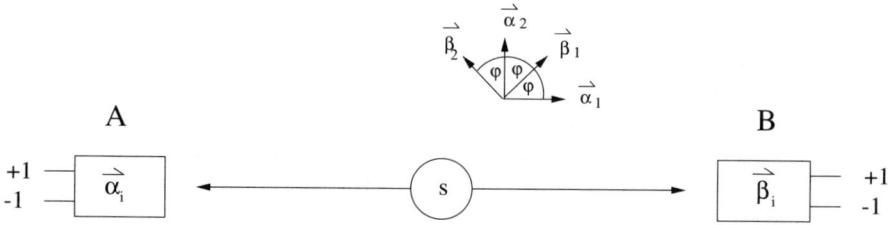

Fig. 1.1. Basic experimental setup for testing Bell's inequality; the choices of the directions of polarization on the Bloch sphere for optimal violation of the CHSH inequality (1.3) correspond to $\varphi = \pi/4$ for spin-1/2 systems

each of these particles are measured subsequently by two distant observers A and B. Observers A and B perform polarization measurements by randomly selecting one of the directions $\boldsymbol{\alpha}_1$ or $\boldsymbol{\alpha}_2$, and $\boldsymbol{\beta}_1$ or $\boldsymbol{\beta}_2$, respectively, in each experiment. Furthermore, let us assume that for each of these directions only two measurement results are possible, namely $+1$ or -1. In the case of photons these measurement results would correspond to horizontal or vertical polarization.

What are the restrictions imposed on correlations of the measurement results if the physical process can be described by an underlying LRT with unknown (hidden) parameters? For this purpose, let us first of all summarize the minimal set of conditions any LRT should fulfill.

1. The state of the two-particle system is determined uniquely by a parameter λ, which may denote an arbitrary set of discrete or continuous labels. Thus the most general observable of observer A or B for the experimental setup depicted in Fig. 1.1 is a function of the variables $(\boldsymbol{\alpha}_i, \boldsymbol{\beta}_j, \lambda)$. If the actual value of the parameter λ is unknown (hidden), the state of the two-particle system has to be described by a normalized probability distribution $P(\lambda)$, i.e. $\int_\Lambda d\lambda\, P(\lambda) = 1$, where Λ characterizes the set of all possible states. The state variable λ determines all results of all possible measurements, irrespective of whether these measurements are performed or not. It represents the element of physical reality inherent in the arguments of EPR: "If, without in any way disturbing a system, we can predict with certainty the value of a physical quantity, then there exists an element of physical reality corresponding to this physical quantity" [24].

2. The measurement results of each of the distant (space-like separated) observers are independent of the choice of polarizations of the other observer. This assumption reflects the locality concept inherent in the arguments of EPR: "The real factual situation of the system A is independent of what is done with the system B, which is spatially separated from the former" [24]. Thus, taking into account also this locality requirement, the most general observable of observer A for the experimental setup depicted in Fig. 1.1 can depend on the variables $\boldsymbol{\alpha}_i$ and λ (for B, $\boldsymbol{\beta}_j$ and λ) only.

These two assumptions, which reflect fundamental notions of classical physics as used in the arguments of EPR, restrict significantly the possible correlations of measurements performed by both distant observers. According to these assumptions, the following measurement results are possible: $a(\boldsymbol{\alpha}_i, \lambda) \equiv a_i = \pm 1$ ($i = 1, 2$) for observer A, and $b(\boldsymbol{\beta}_i, \lambda) \equiv b_i = \pm 1$ ($i = 1, 2$) for observer B. For a given value of the state variable λ, all these possible measurement results of the dichotomic (two-valued) variables a_i and b_i ($i = 1, 2$) can be combined in the single relation

$$|(a_1 + a_2)b_1 + (a_2 - a_1)b_2| = 2 \,. \tag{1.1}$$

It should be mentioned that this relation is counterfactual [26] in the sense that it involves both results of actually performed measurements and possible results of unperformed measurements. All these measurement results are determined uniquely by the state variable λ. If this state variable is unknown (hidden), (1.1) has to be averaged over the corresponding probability distribution $P(\lambda)$. This yields an inequality for the statistical mean values,

$$\langle a_i b_j \rangle_{\text{LRT}} = \int_\Lambda d\lambda\, P(\lambda) a(\boldsymbol{\alpha}_i, \lambda) b(\boldsymbol{\beta}_j, \lambda) \quad (i, j = 1, 2), \tag{1.2}$$

which is a variant of Bell's inequality and which is due to Clauser, Horne, Shimony and Holt (CHSH) [27], namely

$$|\langle a_1 b_1 \rangle_{\text{LRT}} + \langle a_2 b_1 \rangle_{\text{LRT}} + \langle a_2 b_2 \rangle_{\text{LRT}} - \langle a_1 b_2 \rangle_{\text{LRT}}| \leq 2. \tag{1.3}$$

This inequality characterizes the restrictions imposed on the correlations between dichotomic variables of two distant observers within the framework of any LRT. There are other, equivalent forms of Bell's inequality, one of which was proposed by Wigner [28] and will be discussed in Chap. 3.

Quantum mechanical correlations can violate this inequality. For this purpose let us consider, for example, the spin-entangled singlet state

$$|\psi\rangle = \frac{1}{\sqrt{2}}(|+1\rangle_A|-1\rangle_B - |-1\rangle_A|+1\rangle_B), \tag{1.4}$$

where $|\pm 1\rangle_A$ and $|\pm 1\rangle_B$ denote the eigenstates of the Pauli spin operators σ_z^A and σ_z^B, with eigenvalues ± 1. Quantum mechanically, the measurement of the dichotomic polarization variables a_i and b_i is represented by the spin operators $\hat{a}_i = \boldsymbol{\alpha}_i \cdot \boldsymbol{\sigma}^A$ and $\hat{b}_i = \boldsymbol{\beta}_i \cdot \boldsymbol{\sigma}^B$. ($\boldsymbol{\sigma}^A$, for example, denotes the vector of Pauli spin operators referring to observer A, i.e. $\boldsymbol{\sigma}^A = \sum_{i=x,y,z} \sigma_i^A \mathbf{e}_i$, where $\mathbf{e}_i$ are the unit vectors.) The corresponding quantum mechanical correlations entering the CHSH inequality (1.3) are given by

$$\langle \hat{a}_i \hat{b}_j \rangle_{\text{QM}} = \langle \psi | \hat{a}_i \hat{b}_j | \psi \rangle = -\boldsymbol{\alpha}_i \cdot \boldsymbol{\beta}_j. \tag{1.5}$$

Choosing the directions of the polarizations $(\boldsymbol{\alpha}_1, \boldsymbol{\beta}_1)$, $(\boldsymbol{\beta}_1, \boldsymbol{\alpha}_2)$, $(\boldsymbol{\alpha}_2, \boldsymbol{\beta}_2)$ on the Bloch sphere so that they involve an angle of $\pi/4$ (see Fig. 1.1), one finds a maximal violation of inequality (1.3), namely

$$|\langle \hat{a}_1 \hat{b}_1 \rangle_{\text{QM}} + \langle \hat{a}_2 \hat{b}_1 \rangle_{\text{QM}} + \langle \hat{a}_2 \hat{b}_2 \rangle_{\text{QM}} - \langle \hat{a}_1 \hat{b}_2 \rangle_{\text{QM}}| = 2\sqrt{2} > 2. \tag{1.6}$$

Thus, for this entangled state, the quantum mechanical correlations between the measurement results of the distant observers A and B are stronger than any possible correlation within the framework of an LRT. Obviously, these correlations are incompatible with the classical notions of reality and locality of any LRT. It is these peculiar quantum correlations originating from entanglement which have been of central interest in research on the foundations of quantum theory and which are also of central interest for quantum information processing.

So far, numerous experiments testing and supporting violations of Bell's inequality [29–31] have been performed.[2] However, from a strictly logical point of view, the results of all these experiments could still be explained by an LRT, owing to two loopholes, namely the locality and the detection loopholes. The locality loophole concerns violations of the crucial locality assumption underlying the derivation of Bell's inequality. According to this assumption one has to ensure that any signaling between two distant observers A and B is impossible. The recently performed experiment of G. Weihs et al. [31] succeeded in fulfilling this locality requirement by choosing the separation between these observers to be sufficiently large. However, so far all experiments have involved low detection efficiencies, so that in principle the observed correlations which violate Bell's inequality can still be explained by an LRT [32,33]. This latter detection loophole constitutes a major experimental challenge, and it is one of the current experimental aims to close both the detection loophole and the locality loophole simultaneously [34–36].

The concepts of physical reality and locality which lead to inequality (1.3) can also lead to logical contradictions with quantum theory which are not of statistical origin. This becomes particularly apparent when one considers an entangled three-particle state of the form

$$|\psi\rangle_{\text{GHZ}} = \frac{1}{\sqrt{2}}(|+1\rangle_A |+1\rangle_B |+1\rangle_C - |-1\rangle_A |-1\rangle_B |-1\rangle_C)\,, \quad (1.7)$$

a so-called Greenberger–Horne–Zeilinger (GHZ) state [37]. Again $|\pm 1\rangle_A$, $|\pm 1\rangle_B$, and $|\pm 1\rangle_C$ denote the eigenstates of the Pauli spin operators σ_z^A, σ_z^B, and σ_z^C, with eigenvalues ± 1. Similarly to Fig. 1.1, let us assume that the polarization properties of this entangled quantum state are investigated by three distant (space-like separated) observers A, B and C. Each of these observers chooses his or her direction of polarization randomly along either the x or the y axis.

What are the consequences an LRT would predict? As the three observers are space-like separated, the locality assumption implies that a polarization measurement by one of these observers cannot influence the results of the other observers. Following the notation of Fig. 1.1, the possible results of the polarization measurements of observers A, B and C along directions $\boldsymbol{\alpha}_i$, $\boldsymbol{\beta}_j$ and $\boldsymbol{\gamma}_k$ are $a_i = \pm 1$, $b_j = \pm 1$ and $c_k = \pm 1$. Let us now consider four possible coincidence measurements of these three distant observers, with results (a_x, b_x, c_x), (a_x, b_y, c_y), (a_y, b_x, c_y) and (a_y, b_y, c_x). As we are dealing with dichotomic variables, within an LRT the product of all these measurement results is always given by

$$R_{\text{LRT}} = (a_x b_x c_x)(a_x b_y c_y)(a_y b_x c_y)(a_y b_y c_x) = a_x^2 b_x^2 c_x^2 a_y^2 b_y^2 c_y^2 = 1\,. \quad (1.8)$$

What are the corresponding predictions of quantum theory? In quantum theory the variables a_i, b_j and c_k are replaced by the Pauli spin operators

---
[2] For a comprehensive discussion of experiments performed before 1989, see [29].

$\hat{a}_i = \boldsymbol{\alpha}_i \cdot \boldsymbol{\sigma}^{\mathrm{A}}$, $\hat{b}_j = \boldsymbol{\beta}_j \cdot \boldsymbol{\sigma}^{\mathrm{B}}$ and $\hat{c}_k = \boldsymbol{\gamma}_k \cdot \boldsymbol{\sigma}^{\mathrm{C}}$. The GHZ state of (1.7) fulfills the relations

$$\hat{a}_x \hat{b}_x \hat{c}_x |\psi\rangle_{\mathrm{GHZ}} = -|\psi\rangle_{\mathrm{GHZ}},$$

$$\hat{a}_x \hat{b}_y \hat{c}_y |\psi\rangle_{\mathrm{GHZ}} = \hat{a}_y \hat{b}_x \hat{c}_y |\psi\rangle_{\mathrm{GHZ}} = \hat{a}_y \hat{b}_y \hat{c}_x |\psi\rangle_{\mathrm{GHZ}} = |\psi\rangle_{\mathrm{GHZ}}. \qquad (1.9)$$

Therefore the quantum mechanical result for the product of (1.8) is given by

$$R_{\mathrm{QM}}|\psi\rangle_{\mathrm{GHZ}} = (\hat{a}_x \hat{b}_x \hat{c}_x)(\hat{a}_x \hat{b}_y \hat{c}_y)(\hat{a}_y \hat{b}_x \hat{c}_y)(\hat{a}_y \hat{b}_y \hat{c}_x)|\psi\rangle_{\mathrm{GHZ}} = (-1)|\psi\rangle_{\mathrm{GHZ}} \qquad (1.10)$$

and contradicts the corresponding result of an LRT. These peculiar quantum mechanical predictions have recently been observed experimentally [38]. The entanglement inherent in these states offers interesting perspectives on the possibility of distributing quantum information between three parties [39].

1.1.2 Characteristic Quantum Effects for Practical Purposes

According to a suggestion of Feynman [40], quantum systems are not only of interest for their own sake but might also serve specific practical purposes. Simple quantum systems may be used, for example, for simulating other, more complicated quantum systems. This early suggestion of Feynman emphasizes possible practical applications and thus indicates already a change in the attitude towards characteristic quantum phenomena.

In the same spirit, but independently, Wiesner suggested in the 1960s the use of nonorthogonal quantum states for the practical purpose of encoding secret classical information [41].[3] The security of such an encoding procedure is based on a characteristic quantum phenomenon which does not involve entanglement, namely the impossibility of copying (or cloning) nonorthogonal quantum states [4]. This impossibility becomes apparent from the following elementary consideration. Let us imagine a quantum process which is capable of copying two nonorthogonal quantum states, say $|0\rangle$ and $|1\rangle$, with $0 < |\langle 0|1\rangle| < 1$. This process is assumed to perform the transformation

$$|0\rangle|\varphi\rangle|a\rangle \to |0\rangle|0\rangle|a_0\rangle,$$

$$|1\rangle|\varphi\rangle|a\rangle \to |1\rangle|1\rangle|a_1\rangle, \qquad (1.11)$$

where $|\varphi\rangle$ represents the initial quantum state of the (empty) copy and $|a\rangle$, $|a_0\rangle$, $|a_1\rangle$ denote normalized quantum states of an ancilla system. This ancilla system describes the internal states of the copying device. As this copying process has to be unitary, it has to conserve the scalar product between the two input and the two output states. This implies the relation $\langle 0|1\rangle(1 - \langle 0|1\rangle \langle a_0|a_1\rangle) = 0$. This equality can be fulfilled only if either states

---
[3] Though this article was written in the 1960s, it was not published until 1983.

$|0\rangle$ and $|1\rangle$ are orthogonal, i.e. $\langle 0|1\rangle = 0$, or if $\langle 0|1\rangle = 1 = \langle a_0|a_1\rangle$. Both possibilities contradict the original assumption of nonorthogonal, nonidentical initial states. Therefore a quantum process capable of copying nonorthogonal quantum states is impossible. This is an early example of an impossible quantum process.

Soon afterwards, Bennett and Brassard [42] proposed the first quantum protocol (BB84) for secure transmission of a random, secret key using nonorthogonal states of polarized photons for the encoding (see Table 1.1). In the Vernam cipher, such a secret key is used for encoding and decoding messages safely [6,43]. In this latter encoding procedure the message and the secret key are added bit by bit, and in the decoding procedure they are subtracted again. If the random key is secret, the safety of this protocol is guaranteed provided the key is used only once, has the same length as the message and is truly random [44]. Nonorthogonal quantum states can help in transmitting such a random, secret key safely. For this purpose A(lice) sends photons to B(ob) which are polarized randomly either horizontally (+1) or vertically (−1) along two directions of polarization. It is convenient to choose the magnitude of the angle between these two directions of polarization to be $\pi/8$. B(ob) also chooses his polarizers randomly to be polarized along these directions. After A(lice) has sent all photons to B(ob), both communicate to each other their choices of directions of polarization over a public channel. However, the sent or measured polarizations of the photons are kept secret. Whenever they chose the same direction (yes), their measured polarizations are correlated perfectly and they keep the corresponding measured results for their secret key. The other measurement results (no) cannot be used for the key. Provided the transmission channel is ideal, A(lice) and B(ob) can use part of the key for detecting a possible eavesdropper because in this case some of the measurements are not correlated perfectly. In practice, however, the transmission channel is not perfect and A(lice) and B(ob) have to process their raw key further to extract from it a secret key [45]. It took some more

Table 1.1. Part of a possible idealized protocol for transmitting a secret key, according to [42]

A(lice)'s direction i	1	2	1	1	2	1	2	2	1	2	···
A(lice)'s polarization	+1	−1	−1	+1	+1	+1	−1	−1	−1	+1	···
B(ob)'s direction i	2	1	1	2	2	1	2	1	1	2	···
B(ob)'s measured polarization	+1	−1	−1	−1	+1	+1	−1	+1	−1	+1	···
Public test of common direction	No	No	Yes	No	Yes	Yes	Yes	No	Yes	Yes	···
Secret key			−1		+1	+1	−1		−1	+1	···

years to realize that an exchange of secret keys can be achieved with the help of entangled quantum states [46]. Thereby, the characteristic quantum correlations of entangled states and the very fact that they are incompatible with any LRT can be used for ensuring security of the key exchange. After the first proof-of-principle experiments [47,48], the first practical implementation of quantum cryptography over a distance of about 1 km was realized at the University of Geneva using single, polarized photons transmitted through an optical fiber [49]. These developments launched the whole new field of quantum cryptography. Now, this field represents the most developed part of quantum information processing. Quantum cryptography based on the BB84 protocol has already been realized over a distance of 23 km [50]. Recent experiments [30,31] have demonstrated that photon pairs can also be entangled over large distances, so that entanglement-based quantum cryptography over such large distances might become accessible soon. Some of these experiments are discussed in Chap. 3.

Simultaneously with these developments in quantum cryptography, numerous other physical processes were discovered which were either enabled by entanglement or in which entanglement led to an improvement of performance. The most prominent examples are dense coding [51], entanglement-assisted teleportation [10,11,52] and entanglement swapping [52,53]. (These processes are discussed in detail in Chaps. 2 and 3.) In the spirit of Feynman's suggestion, all these developments demonstrate that characteristic quantum phenomena have practical applications in quantum information processing.

1.1.3 Quantum Algorithms

Feynman's suggestion also indicates interesting links between quantum physics and computer science. After the demonstration [54] that quantum systems can simulate reversible Turing machines [55], the first quantum generalization of Turing machines was developed [7]. (Turing machines are general models of computing devices and will be discussed in detail in Chap. 4.) Furthermore, it was pointed out that one of the remarkable properties of such a quantum Turing machine is quantum parallelism, by which certain tasks may be performed faster than by any classical computing device. Deutsch's algorithm [7,56] was the first quantum algorithm demonstrating how the interplay between quantum interference, entanglement and the quantum mechanical measurement process could serve this practical purpose.

The computational problem solved by Deutsch's algorithm is the following. We are given a device, a so-called oracle, which computes a Boolean function f mapping all possible binary n-bit strings onto one single bit. Therefore, given a binary n-bit string x as input, this oracle can compute $f(x) \in \{0,1\}$ in a single step. Furthermore, let us assume that this function is either constant or balanced. Thus, in the first case the 2^n possible input values of x are all mapped onto 0 or all onto 1. In the second case half of the input values are mapped onto either 0 or 1 and the remaining half are mapped onto the other value. The problem is to develop an algorithm which determines whether f is constant or balanced.

Let us first of all discuss briefly the classical complexity of this problem. In order to answer the question in the worst possible case, the oracle has to be queried more than 2^{n-1} times. It can happen, for example, that the first 2^{n-1} queries all give the same result, so that at least one more query of the oracle is required to decide whether f is constant or balanced. Thus, classically, it is apparent that the number of steps required grows exponentially with the number of bits.

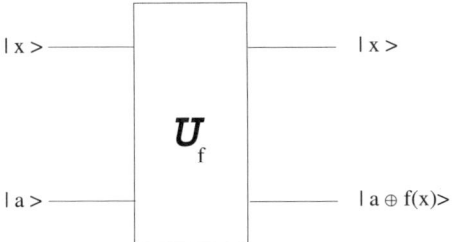

Fig. 1.2. Basic operation of a quantum oracle $\mathcal{U}_f$ which evaluates a Boolean function $f : x \in \mathbf{Z}_2^n \to f(x) \in \mathbf{Z}_2^1 \equiv \{0,1\}$; $|x\rangle$ is the input state of an n-qubit quantum system; $|a\rangle$ is a one-qubit state and $\oplus$ denotes addition modulo 2

Quantum mechanically, the situation is different. The 2^n possible binary n-bit strings x can be represented by quantum states $|x\rangle$, which form a basis in a 2^n-dimensional Hilbert space $\mathcal{H}_{2^n}$, which is the state space of n qubits. Furthermore, we imagine that the classical oracle is replaced by a corresponding quantum oracle (Fig. 1.2). This is a unitary transformation $\mathcal{U}_f$ which maps basis states of the form $|x\rangle|a\rangle$, where $a \in \{0,1\}$, to output states of the form $|x\rangle|a \oplus f(x)\rangle$ in a single step. Here, $|a\rangle$ denotes the quantum state of an ancilla qubit and $\oplus$ denotes addition modulo 2. If the initial state is $|x\rangle|0\rangle$, for example, the quantum oracle performs an evaluation of $f(x)$, resulting in the final state $|x\rangle|f(x)\rangle$. However, as this transformation is unitary, it can perform this task also for any linear combination of possible basis states in a single step. This is the key idea of quantum parallelism [7]. Deutsch's quantum algorithm obtains the solution to the problem posed above by the following steps (Fig. 1.3):

1. The n-qubit quantum system and the ancilla system are prepared in states $|0\rangle$ and $(|0\rangle - |1\rangle)/\sqrt{2}$. Then a Hadamard transformation

$$H : |0\rangle \to \frac{1}{\sqrt{2}}(|0\rangle + |1\rangle) ,$$
$$|1\rangle \to \frac{1}{\sqrt{2}}(|0\rangle - |1\rangle) \qquad (1.12)$$

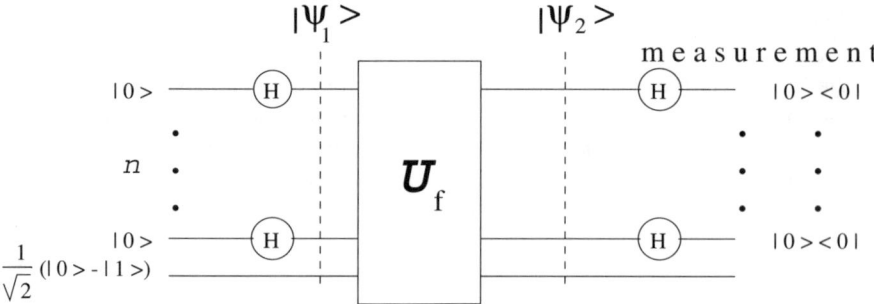

Fig. 1.3. Schematic representation of Deutsch's quantum algorithm

is applied to all of the first n qubits. We denote by $H^{(i)}$ the application of H to the ith qubit. Thus, the separable quantum state

$$|\psi_1\rangle \equiv \frac{1}{\sqrt{2}}[(\prod_{i=1}^{n} \otimes H^{(i)})|0\rangle](|0\rangle - |1\rangle) = \frac{1}{\sqrt{2^{n+1}}} \sum_{x \in 2^n} |x\rangle(|0\rangle - |1\rangle) \quad (1.13)$$

is prepared.

2. A single application of the quantum oracle $\mathcal{U}_f$ to state $|\psi_1\rangle$ yields the quantum state

$$|\psi_2\rangle \equiv \mathcal{U}_f|\psi_1\rangle = \frac{1}{\sqrt{2^{n+1}}} \sum_{x \in 2^n} (-1)^{f(x)}|x\rangle(|0\rangle - |1\rangle) \ . \quad (1.14)$$

3. Subsequently a quantum measurement is performed to determine whether the system is in state $|\psi_1\rangle$ or not. With the help of n Hadamard transformations (as in step 1), this quantum measurement can be reduced to a measurement of whether the first n qubits of the quantum system are in state $|0\rangle$ or not.

If in step 3 the quantum system is found in state $|\psi_1\rangle$, f is constant, otherwise f is balanced. One of these two possibilities is observed with unit probability. The probability p of observing the quantum system in state $|\psi_1\rangle$ is given by

$$p \equiv |\langle\psi_1|\psi_2\rangle|^2 = \frac{1}{2^n} |\sum_{x \in 2^n} (-1)^{f(x)}|^2 \ . \quad (1.15)$$

Taking into account the single application of the quantum oracle in step 2 and the application of the Hadamard transformations in the preparation and measurement processes, Deutsch's quantum algorithm requires $O(n)$ steps to obtain the final answer, in contrast to any classical algorithm, which needs an exponential number of steps. Thus Deutsch's quantum algorithm leads to an exponential speedup.

A key element of this quantum algorithm and of those discovered later is the quantum parallelism involved in step 2, where the linear superposition of the first n qubits comprises the requested global information about the function f. For most of the possible functions f this intermediate quantum state is expected to be entangled. An exception is the case of a constant function f, for which the quantum state $|\psi_2\rangle$ is separable. Furthermore, it is also crucial for the success of this quantum algorithm that the final measurement in step 3 yielding the required answer can be implemented by a fast quantum measurement whose complexity is polynomial in n. This is a requirement fulfilled by all other known fast quantum algorithms. The quantum algorithm described above was the first example demonstrating that quantum phenomena may speed up computations in such a way that an exponential gap appears between the complexity class of the quantum problem and the complexity class of the corresponding classical probabilistic problem.

Continuing this development initiated by Deutsch, other, new fast quantum algorithms were discovered in the subsequent years. The most prominent examples are Simon's quantum algorithm [57], Shor's celebrated algorithm [9] for factorizing numbers, and Grover's search algorithm [58]. (Quantum algorithms are discussed in detail in Chap. 4.) In addition, possible realizations of quantum computing devices were suggested which were based on trapped ions [59] and on cavity quantum electrodynamical setups [60]. These developments called for new methods for stabilizing quantum algorithms against perturbing environmental influences, which tend to destroy quantum interference and quantum entanglement [61]. This led to the development of the first error-correcting codes [62–66] by adaptation of classical error-correcting techniques to the quantum domain. An introduction to the theory of quantum error correction is presented in Chap. 4.

1.2 Quantum Physics and Information Processing

What are the common features of these early developments? The common element of these early developments in quantum cryptography and quantum computation is that they all involve the practical processing of information and they are all founded on and facilitated by characteristic quantum phenomena. These phenomena, among which the most prominent is entanglement, are in conflict with the classical concepts of physical reality and locality. Obviously, these early developments hint at a profound connection between the concept of information and some fundamental concepts of quantum theory, which is also promising from the technological point of view. It is these technologically oriented aspects of quantum information theory [67–69] which are at the heart of quantum information processing.

Methods for processing quantum information have developed rapidly during the last few years [12]. Owing to significant experimental advances, basic interference and entanglement phenomena which are of central interest for processing quantum information have been realized in the laboratory in various physical systems. Basic schemes for quantum communication have been demonstrated with photons [10,11,49,70]. Realizations of elementary quantum logical operations have been based on trapped ions [13,14] and on nuclear magnetic resonance [15]. Recent experiments indicate that besides cavity quantum electrodynamical setups [16], trapped neutral atoms which are guided along magnetic wires (atom chips) might also be useful for quantum information processing [17]. There have also been theoretical proposals on using ultracold atoms in optical lattices [18,19], on ions in an array of microtraps [20] and on solid-state devices [21–23] for the implementation of quantum logical gates.

By now, quantum information processing has become an interdisciplinary subject which attracts not only physicists but also researchers from other communities. The common interest is the practical, technologically oriented application of characteristic quantum phenomena. At this stage of development, it appears necessary to examine recent achievements and to emphasize the underlying, general, basic concepts, which have been developing gradually and which are now commonly adopted by all researchers in this field. This is one of the main intentions of the rest of the book.

In Chap. 2, Werner introduces the basic concepts of quantum information theory and describes the fundamental mathematical structures underlying recent and current developments. In particular, this chapter addresses a natural question appearing in connection with Feynman's suggestion, namely what can be done with the help of quantum systems and what cannot be done. A first example of an impossible quantum process, the copying of nonorthogonal quantum states, has already been mentioned. Other examples of possible and impossible quantum processes are discussed in detail in this contribution.

First experimental realizations of basic quantum communication schemes based on entangled photon pairs are discussed in Chap. 3 by Weinfurter and Zeilinger. These first experiments on entanglement-based quantum cryptography, dense coding and quantum teleportation demonstrate the important role photons play in current experiments. Furthermore, these experiments also emphasize once again the fundamental significance of entanglement for quantum information processing.

The basic theoretical concepts of quantum computation and the mathematical structure underlying quantum algorithms are discussed in Chap. 4 by Beth and Rötteler. In particular, it is demonstrated how recent results in the theory of signal processing can be used for the development of new fast quantum algorithms. A short introduction to the theory of quantum error correction is also presented.

A comprehensive account of the mathematical structure of entanglement and of the significance of mixed entangled states for quantum information processing is presented in Chap. 5 by M. Horodecki, P. Horodecki and R. Horodecki. One of the most surprising recent developments in this context has been the discovery of bound entanglement [71]. Though much is still unknown, this section gives a state-of-the-art presentation of what is known about this new form of entanglement and its implications for processing quantum information.

2 Quantum Information Theory
– an Invitation

Reinhard F. Werner

2.1 Introduction

Quantum information and quantum computers have received a lot of public attention recently. Quantum computers have been advertised as a kind of warp drive for computing, and indeed the promise of the algorithms of Shor and Grover is to perform computations which are extremely hard or even provably impossible on any merely "classical" computer. On the experimental side, perhaps the most remarkable feat of quantum information processing was the realization of "quantum teleportation", which once again has science fiction overtones.

In some sense these miracles are an extension of the strangeness of quantum mechanics – those unresolved questions in the foundations of quantum mechanics, which most physicists know about, but few try to tackle directly in their research. However, trying to build an explanation of quantum information on the literature about the foundations of quantum mechanics is more likely to mystify than to clarify. It would also give a wrong idea of how discussions in this new field are conducted. Because, just as physicists with widely differing convictions on foundational matters can usually agree quite easily on what the predictions of quantum mechanics are in a particular experimental setup, researchers in quantum information can agree on whether a device should work, no matter what they may think about the deeper meaning of the wave function. For example, one of the founders of the field is an outspoken proponent of the many-worlds interpretation of quantum mechanics (which I, personally, find useless and bizarre). But, whatever the intuitions leading him to his discoveries about quantum computing may have been, these discoveries make sense in every other interpretation.

In this article I shall give an account of the basic concepts of quantum information theory, staying as much as possible in the area of general agreement. So, in order to enter this new field, plain quantum mechanics is enough, and no new, perhaps obscure, views are needed. There is, of course, a characteristic shift in emphasis expressed by the word "information", and we shall have to explore the consequences of this shift.

The article is divided into two parts. The first (up to the end of Sect. 2.5) is mostly in plain English, centered around the exploration of what can or cannot be done with quantum systems as information carriers. The second part, Sect. 2.6, then gives a description of the mathematical structures and of some of the tools needed to develop the theory.

2.2 What is Quantum Information?

Let us start with a preliminary definition:

Quantum information is that kind of information which is carried by quantum systems from the preparation device to the measuring apparatus in a quantum mechanical experiment.

So a "transmitter" of quantum information is nothing but a device preparing quantum particles, and a "receiver" is just a measuring device. Of course, this is not saying much. But even so, it is a strange statement from the point of view of classical information theory: in that theory one usually does not care about the physical carrier of the information, or else one would have to distinguish "electrodynamical information", "printed information", "magnetic information" and many more. In fact, the success of (classical) information theory depends largely on abstracting from the physical carrier, and going instead for the general principles underlying any information exchange. So why should "quantum information" be any different?

A moment's reflection makes clear why the abstraction from the physical carrier of information leads to a successful theory: the reason is that it is so easy to convert information between all such carriers. The conversion from bytes on a hard disk, to currents in a chip, to signals on a cable, to radio waves via satellite and maybe, finally, to an image on a computer screen in another continent all happens essentially without loss, and if there are losses, they are well understood, and it is known how to correct for them. Therefore, the crucial question is: can "quantum information" in the above loose sense also be converted to those standard classical kinds of information, and back, without loss? Or: are there fundamental limitations to such a translation, and is quantum information hence really a *new* kind of information?

This book would not have been written if the answer to the last question were not affirmative: quantum information is indeed a new kind of information. But to make this precise, let us see what would be required of a successful translation. Let us begin with the conversion of quantum information to classical information: a device for this conversion would take a quantum system and produce as its output some classical information. This is nothing but a complicated way of saying "measurement". The reverse translation, from classical to quantum information, obviously involves some preparation of quantum systems. The classical input to such a device is used to control the settings of the preparing device, and any dependence of the preparation process on classical information is admissible. There are two kinds of devices we can combine from these two elements. Let us first consider a device going from classical to quantum to classical information. This is a rather

Fig. 2.1. Classical teleportation. Here and in the following diagrams, a *wavy arrow* stands for quantum systems, and a *straight arrow* for the flow of classical information

commonplace operation. For example, one can encode one classical bit in the polarization degree of freedom of a photon (clearly a quantum system), by choosing one of two orthogonal polarizations for the photon, depending on the value of the classical bit. The readout is done by a photomultiplier combined with a polarization filter in one of the corresponding directions. In principle, this allows a perfect transmission. In some sense every transmission of classical information is of this kind, because every physical system ultimately obeys the laws of quantum mechanics, even if we can often disregard this fact and treat it classically. Hence classical information can be translated into quantum information (and back).

But what about the converse? This hypothetical (and in fact, impossible) process has come to be known as *classical teleportation* (see Fig. 2.1). It would involve a measuring device M, operating on some input quantum systems. The results of the measurements are subsequently fed into a preparing device P, which produces the final output of the combined device. The task is to set things up such that the outputs of the combined device are indistinguishable from the quantum inputs. Of course, we have to say precisely what "indistinguishable" should mean. Clearly, this cannot mean that "the same" system comes out at the other end. In the classical case this is not demanded either. What can only be meant in quantum mechanics is that *no statistical test will see the difference*. In other words, no matter what the preparation of the input systems is and no matter what observable we measure on the outputs of the teleportation device, we shall always get the same probability distribution of results as if the inputs had been directly measured. Note also that this criterion does not involve the states of individual systems, but only states in the form of the distribution parameters of ensembles of identically prepared systems.

The impossibility of classical teleportation will be treated extensively in the following section, where it is related to a hierarchy of impossible machines. For a mathematical statement of this impossibility in the standard quantum formalism of quantum mechanics, see the remark after (2.7). For the moment, however, let us take it for granted, and see what all this says about the new concept of quantum information.

First of all, we are concerned here with problems of transmission, not with content or meaning. This is exactly the same as in classical information theory. There, too, it is often not easy to avoid confusion with a different concept of "information" used in everyday language, namely the kind available at an information desk. Information theory does not care whether a TV channel is used for "misinformation", but can say everything about what it takes to ensure the technical quality of the final images. Hence the quantitative measures of "information" all relate to storage and transmission capacity, to the possibilities of compression and error correction and so on. In the same vein, quantum information theory will not tell us what the meaning of a "quantum message" is, and this is probably meaningless anyway, because a message that has been "read" is classical almost by definition. But quantum information theory has precise notions of the resources needed to transmit such information faithfully.

Secondly, transmission of quantum information is not at all an exotic concept in the context of modern physics. It can be paraphrased in various, perhaps more familiar ways, for example as "transmission of intact quantum states", as "coherent transmission of quantum systems" or as transmission "preserving all interference possibilities" of the system. Nevertheless the information metaphor is useful, not only because it suggests new applications, but also because it leads one to ask new questions, and leads to quantitative notions where previously there was only a qualitative understanding. And possibly this even provides a way to see in a sharper light the old conundrums of the foundations of quantum mechanics.

2.3 Impossible Machines

The usefulness of considering impossible machines is well known from thermodynamics: the second law of thermodynamics is often stated as the impossibility of a perpetual-motion machine. The theorem of the impossibility of classical teleportation is likewise a fundamental law of quantum mechanics, and a lot can be learned from analyzing it. Typically, the impossible machines of quantum theory are perfectly possible in classical physics, so their impossibility does not follow superficially from their description, but rather carries a connotation of paradox.

We shall discuss a range of impossible tasks, consisting of

- teleportation
- copying ("cloning")
- joint measurement
- Bell's telephone.

As we shall see, teleportation is the most powerful of these, in the sense that if we had a teleportation device, we could build a quantum copier, from which we could in turn construct joint measurements and, finally, a device known as Bell's Telephone, by which we could set up superluminal communication. Hence, if we uphold the principle of causality, which forbids the weakest machine in this hierarchy, we are certain that teleportation is likewise impossible. In this section we shall follow this line of reasoning to prove the impossibility of teleportation. Of course, there are other, more direct ways of proving this result from the structure of quantum mechanics. However, these usually require more of the quantum formalism and give less insight into the differences between classical and quantum information.

2.3.1 The Quantum Copier

This is the machine referred to in the well-known paper of Wootters and Zurek entitled "A single quantum cannot be cloned" [4]. By definition, a copier would be a device taking one quantum system as input and turning out two systems of the same type. The condition for calling this a (faithful) copier is that we would not be able to distinguish a system coming from the output from the input system by any statistical test, i.e. by means of the probabilities measured for any observable, and for any preparation of the initial state. Hence the device has to operate on arbitrary "unknown" states. It is clear that a copier in the ordinary sense, e.g. a mail relay distributing email to several recipients, indeed satisfies this condition in the domain of classical information. Note that we are not so unreasonable as to demand what the paper quoted above suggests, namely that we could test this device on *single events*, or even assume some ontological "identity" of input and output: the criterion for faithful copying is flatly statistical, and can be verified by a straightforward collection of statistical tests.

Given a teleportation device, building a copier is quite easy (see Fig. 2.2). All we have to do is to remember that the classical information obtained in the intermediate stage of the teleportation process can be copied perfectly. Hence we can apply the measuring device of the teleportation line to the input system, copy the results, and simply run the reconstructing preparation process on each of these copies.

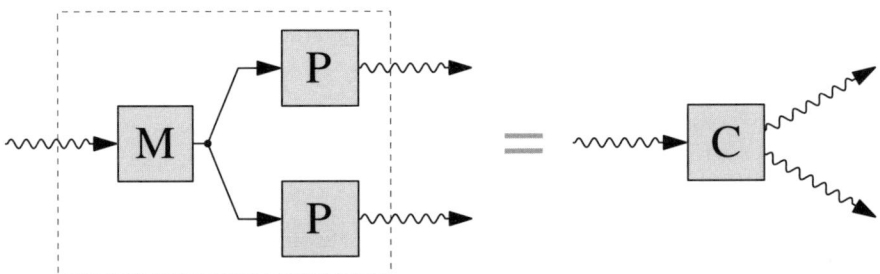

Fig. 2.2. Making a copier from a "classical teleportation" line

2.3.2 Joint Measurement

This is the task of combining two separate measuring devices into a single device, or the "simultaneous measurement" of two quantum observables A and B. Thus, a joint measuring device "$A\&B$" is a device giving a pair (a,b) of classical outputs each time it is operated, such that a is a possible output of A, and b is a possible output of B. (We use the symbol A to denote both an observable and a device that measures this observable, and similar for B.) We require that the statistics of the a outcomes alone are the same as for device A, and similarly for B. Note that once again our criterion is statistical, and can be tested without recourse to counterfactual conditionals such as "the result which would have resulted if B rather than A had been measured on this particular quantum particle".

Many quantum observables are not jointly measurable in this sense. The most famous examples, position and momentum, different components of angular momentum, and positions of a free particle at different times, are probably contained in every quantum mechanics course. Hence the impossibility of joint measurements is nothing but a precise statement of an aspect of "complementarity".

Nevertheless, a joint measurement device for any of these could readily be constructed given a functioning quantum copier (see Fig. 2.3): one would simply run the copier C on the quantum system, and then apply the two given measuring devices, A and B, to the copies. It is easy to see that the definition of the copier then guarantees that the statistics of a and b separately come out right. In other words, a copier can be seen as a *universal joint measuring device*.

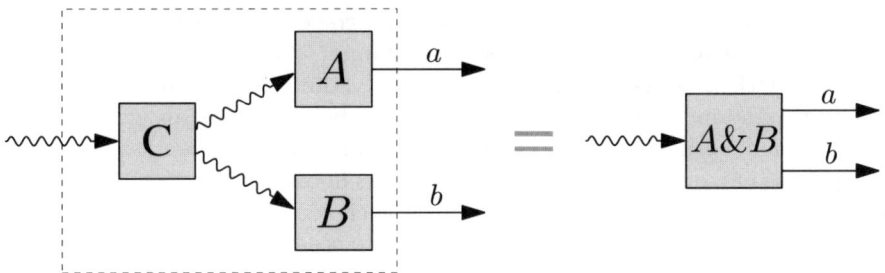

Fig. 2.3. Obtaining joint measurements from a copier

2.3.3 Bell's Telephone

This is not named after a certain phone company, but after John S. Bell, who never proposed it in this form, but might have. It refers to a project of performing superluminal communication using only correlations of the type tested by Bell's inequalities. Without going into details for the moment, the basic setup would consist of a source producing pairs of particles and sending one member of each pair to each of the two communicating parties, conventionally named "Alice" and "Bob". Each of them has a collection of different measuring devices to choose from, and the idea is for Alice to do something which creates a noticeable change in the probabilities measured by Bob. Clearly, this is a paradoxical task, because no particle or other physical carrier of information actually goes from Alice to Bob. Therefore, if the particles move sufficiently far apart from one another, this device transmits superluminally.

It is maybe useful to point out here a common confusion concerning such superluminal effects, which sometimes even afflicts otherwise reliable professional writers. The mistake can usually be spotted easily by a device I call the "ping-pong ball test". It goes like this:

> Take an author's explanation of Bell's inequalities, and substitute "ping-pong balls" for every quantum particle. Then if whatever the author is selling as paradoxical remains true, he/she hasn't understood a thing.

Here is an example: imagine a box containing a ping-pong ball; the box can be separated into two parts, without anyone looking at the ball. One part is shipped to Tokyo or Alpha Centauri, without anyone looking inside. Then if I open the other box I know instantly, i.e. "at superluminal speed", whether the ball is at the distant location or not. Of course, this is true but hardly paradoxical, and is totally useless for sending a message either way. To repeat: there is nothing paradoxical in statistical correlations per se between distant systems with a common past, even if the correlation is perfect.

If Alice wants to send a message to Bob, correlations between two measuring devices are useless, because they cannot even be detected without comparing the results, which requires exactly the communication the Telephone was intended for. Only if something Alice *does* has an effect on the measurement results at Bob's end can we speak of communication. The only thing Alice can do in the standard setup is to choose a measuring device, and Bell's Telephone can be said to work if these choices have an influence on the probabilities measured by Bob (who has no access to Alice's measurement results). If there is no physical system traveling from Alice to Bob, however, this will be impossible.

To be fair, this can hardly be counted as an impossible machine of quantum mechanics, since the argument has nothing to do with quantum theory. What makes it fit into the hierarchy described here is the following: if we assume that Bob has a joint measuring device for two yes/no measurements, and Bell's inequalities are violated, we can design a strategy for Alice to send signals to Bob with better than chance results. Hence the joint measurement of suitable observables can provide a device sufficiently strong to achieve a

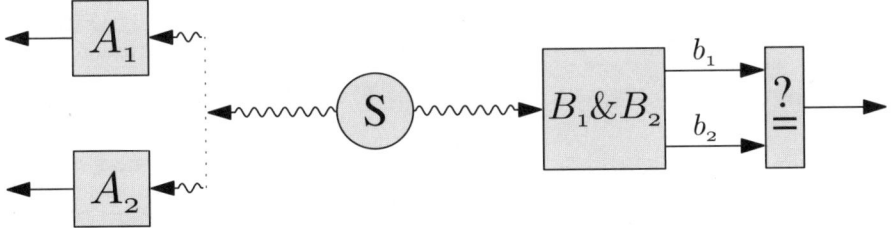

Fig. 2.4. Building Bell's Telephone from a joint measurement

task forbidden by causality, and hence is impossible in general. This is the link between the last two elements in the hierarchy of impossible machines mentioned at the beginning of Sect. 2.3.

The proof of this step amounts to yet another derivation of Bell's inequalities, but since it emphasizes the communication aspect it fits well into our context, and we shall at least sketch it. This step will be rather more technical than the rest of this section, but does not require any quantum theory. The argument can be skipped without loss as far as later sections are concerned.

So let us assume that Alice and Bob each have at their disposal two measuring devices, say A_1, A_2 and B_1, B_2, respectively. Each of these can give a result of either $+1$ or -1. We shall denote by $\mathbb{P}(a,b \mid A_i, B_j)$ the probability for Alice to obtain a and Bob to obtain b in a correlation experiment in which Alice uses measuring device A_i and Bob uses B_j. By

$$C(A_i, B_j) = \sum_{a,b} ab\, \mathbb{P}(a,b \mid A_i, B_j)$$

we shall denote the correlation coefficient, which lies between -1 and $+1$. The combination

$$\beta = C(A_1, B_1) + C(A_1, B_2) + C(A_2, B_1) - C(A_2, B_2) \tag{2.1}$$

carries special significance, as we shall see below. Because the inequality $\beta \leq 2$ is known as the Bell inequality (see Sect. 1.1.1.), we shall call β the *Bell correlation* for this choice of four observables. It is a quantity directly accessible to experiment. Note that Bob usually cannot tell from his data which apparatus (A_1 or A_2) Alice chose. This is reflected by the equation

$$\sum_a \mathbb{P}(a,b \mid A_1, B_j) = \sum_a \mathbb{P}(a,b \mid A_2, B_j) \equiv \mathbb{P}(b \mid B_j),$$

and is borne out by all known experimental data. Now suppose Bob has a joint measuring device for his B_1 and B_2, which we shall denote by $B_1 \& B_2$, which produces pair outcomes (b_1, b_2) (see Fig. 2.4). We can then determine the probabilities $p_i(a_i, b_1, b_2) = \mathbb{P}(a_i, (b_1, b_2) \mid A_i, B_1 \& B_2)$. The condition that this is really a joint measurement is expressed by the equations

$$\sum_{b_1} p_i(a_i, b_1, b_2) = \mathbb{P}(a_i, b_2 \mid A_i, B_2) \quad \text{and} \tag{2.2}$$

$$\sum_{b_2} p_i(a_i, b_1, b_2) = \mathbb{P}(a_i, b_1 \mid A_i, B_1) \,, \tag{2.3}$$

each for $i = 1, 2$. The basic rule for the information transmission is the following:

> Alice encodes the bit she wants to send by choosing either apparatus A_1 or apparatus A_2. Then Bob looks at his readout and interprets it as "A_1" whenever the two displays coincide ($b_1 = b_2$), and as "A_2" if they are different.

We can then estimate the probability p_{ok} for Bob to be right, assuming that the choices A_1 and A_2 are made with the same frequency. Assume first that Alice chooses A_1. Then Bob is right with probability

$$\sum_{a_1, b_1, b_2} \left| \frac{b_1 + b_2}{2} \right| |a_1| \, p_1(a_1, b_1, b_2) \,,$$

where the first factor takes into account the condition $b_1 = b_2$, and the second is introduced for later convenience. Combining this with a second term of similar kind for Alice's choice A_2, and taking into account the probability $1/2$ for each of these choices, we obtain the overall probability p_{ok} for Bob to be correct as

$$\begin{aligned}
p_{\text{ok}} &= \frac{1}{2} \sum_{a_1, b_1, b_2} \left| \frac{b_1 + b_2}{2} \right| |a_1| \, p_1(a_1, b_1, b_2) \\
&\quad + \frac{1}{2} \sum_{a_2, b_1, b_2} \left| \frac{b_1 - b_2}{2} \right| |a_2| \, p_2(a_2, b_1, b_2) \\
&\geq \frac{1}{4} \sum_{a_1, b_1, b_2} (b_1 + b_2) a_1 \, p_1(a_1, b_1, b_2) \\
&\quad + \frac{1}{4} \sum_{a_2, b_1, b_2} (b_1 - b_2) a_1 \, p_2(a_2, b_1, b_2) \\
&= \frac{1}{4} \Big(C(A_1, B_1) + C(A_1, B_2) + C(A_2, B_1) - C(A_2, B_2) \Big) \\
&= \frac{\beta}{4} \,.
\end{aligned} \tag{2.4}$$

Bob is right with a better probability than chance if $p_{\text{ok}} > 1/2$, which, by this computation, can be guaranteed if $\beta > 2$, i.e. if the classical Bell inequality (in Clauser–Horne–Shimony–Holt form [72]) is violated. But this is indeed the case in experiments conducted to determine β (e.g. [73]), which give roughly $\beta \approx 2\sqrt{2} \approx 2.8$. If we believe these experiments, the only conclusion can be that the joint measurability of the B_1 and B_2 used in the experiment would be sufficient to make Bell's Telephone work, which was our claim.

2.3.4 Entanglement, Mixed-State Analyzers and Correlation Resolvers

Violations of Bell's inequalities can also be seen to prove the existence of a new class of correlations between quantum systems, known as *entanglement*. This concept is as fundamental to the field of quantum information theory as the idea of quantum information itself. So rather than organizing this introduction as an answer to the the question "why quantum information is different from classical information", we could have followed the line "why entanglement is different from classical correlation". There are impossible machines in this line of approach, too, and we shall now describe briefly how they fit in.

Consider a correlation experiment of the kind used in the study of Bell's inequalities (see Sect. 2.3.3). If Bob looks at his particles, and makes measurements on them without any communication from Alice, he will find that their statistics are described by a certain mixed state. The state must be mixed, because if he now listens to Alice and sorts his particles according to Alice's measurement results, he will get two subensembles, which are in general different. In the usual ideal 2-qubit situation, in which one obtains the maximal violation of Bell's inequalities, these subensembles are described by pure states.

This is very satisfying for people who see the occurrence of mixed states in quantum mechanics merely as a result of ignorance, as opposed to the deeper kind of randomness encoded in pure states. This view usually comes with an *individual-state interpretation* of quantum mechanics, in which each individual system can be assigned a pure state (a single vector in Hilbert space), and a general preparation procedure is given not just by its density matrix, but by a specific probability distribution of pure states. Let us use the term *mixed-state analyzer* for a hypothetical device which can see the difference, i.e. a measuring device whose output after many measurements on a given ensemble is not just a collection of expectations of quantum observables, but the distribution of pure states in the ensemble. In the case of a correlation experiment, where Bob sees a mixed state only because he is ignorant about Alice's results, this machine would find for him the decomposition of his mixed state into two pure states.

The problem is, of course, that Alice has several choices of measuring devices, and that the decomposition of Bob's mixed state depends, accordingly, on Alice's choice. Hence she could signal to Bob, and we would have another instance of Bell's Telephone. There would be a way out if joint measurements were available (to Alice in this case): then we could say that the two decompositions were just the first step in an even finer decomposition, a further reduction of ignorance, which would be brought to light if Alice were to apply her joint measurement. Presumably the mixed-state analyzer would then yield this finer decomposition, because the operation of this device would not depend on how closely Alice cared to look at her particles.

But just as two quantum observables are often not jointly measurable, two decompositions of mixed states often have no common refinement (actually, in the formalism of quantum theory, these are two variants of the same theorem). In particular, the two decompositions belonging to Alice's choices in an experiment demonstrating a violation of Bell's inequalities have no common refinement, and any mixed-state analyzer could be used for superluminal communication in this situation.

Another device, which is suggested by the individual-state interpretation, arises from a naive extrapolation of this view to the parts of a composite system: if every single system could be assigned a pure state, a composite system could be assigned a pair of pure states, one for each subsystem. A correlated state should therefore be given by a probability distribution of such pairs. A device which represented an arbitrary state of a composite system as a mixture of uncorrelated pure product states might be called a *correlation resolver*. It could be built given a classical teleportation line: when one applies teleportation to one of the subsystems and applies conditions on the classical measurement results of the intermediate stage, one obtains precisely a representation of an arbitrary state in this form. But it is easy to see that any state which can be so analyzed automatically satisfies all Bell-type inequalities, and hence once again the experimental violations of Bell's inequalities show that such a correlation resolver cannot exist. Hence we have here a second line of reasoning in favor of the no-teleportation theorem: a teleportation device would allow classical correlation resolution, which is shown to be impossible by the Bell experiments.

The distinction between resolvable states and their complement is one of the starting points of entanglement theory, where the "resolvable" states are called "separable", or "classically correlated", and all others are called "entangled". For a more detailed treatment and an up-to date overview, the reader is referred to Chap. 5.

Without going into philosophical discussions about the foundations of quantum mechanics, I should like to comment briefly on the individual-state interpretation, which has suggested the two impossible machines discussed in this subsection. First, this view is not at all uncommon, and it is quite possible to read some passages from the masters of the Copenhagen interpretation as an endorsement of this view. Secondly, if we define a *hidden-variable theory* as a theory in which individual systems are described by classical parameters, whose distribution is responsible for the randomness seen in quantum experiments, we have no choice but to call the individual-state interpretation a hidden-variable theory. The hidden variable in this theory is usually denoted by ψ. And sure enough, as we have just pointed out, this theory has all the difficulties with locality such a theory is known to have on general grounds. Thirdly, avoiding an individual-state interpretation, and with it some of its misleading intuitions, is easy enough. In practice this is done anyhow, by concentrating on those aspects of the theory which have some direct statistical meaning, and not on these involving hypothetical, and usually impossible devices. This common ground is the statistical interpretation of quantum mechanics, in which states (pure or mixed) are the analogues of classical probability distributions, and are not seen as a property of an individual system, but of a specific way of preparing the system.

2.4 Possible Machines

2.4.1 Operations on Multiple Inputs

The no-teleportation theorem derived in the previous section says that there is no way to measure a quantum state in such a way that the measuring results suffice to reconstruct the state. At first sight this seems to deny that the notion of "quantum states" has an operational meaning at all. But there is no contradiction, and we shall resolve the apparent conflict in this subsection, if only to sharpen the statement of the no-teleportation theorem.

Let us recall the operational definition of quantum states, according to the statistical interpretation of quantum mechanics. A state is a description of a way of preparing quantum systems, and in all its aspects it is related to computing expectation values. We might also say that it *is* the assignment of an expectation value to every observable of the system. So to the extent that expectation values can be measured, it is possible to determine the state by testing it on sufficiently many observables. What is crucial, however, is that even the determination of a single expectation value is a *statistical* measurement. Hence such a determination requires a repetition of the experiment many times, using many systems prepared according to the same procedure. In contrast, the above description of teleportation demands that it works with a single quantum system as input, and that the measuring device does not accumulate results from several input systems. Expressed in the current jargon, teleportation is required to be a *one-shot operation*. Note that this does not contradict our statistical criteria for the success of teleportation and of other devices, which involve a statistics of independent "single shots".

If we have available many identically prepared systems, many operations which are otherwise impossible become easy. Let us begin with classical teleportation. Its multiinput analogue is the *state estimation* problem: how can we design a measurement operating on samples of many (say, N) systems from the same preparing device, such that the measurement result in each case is a collection of classical parameters forming a Hermitian matrix which on average is close to the density matrix describing the initial preparation.

This is symbolized in Fig. 2.5 (with the box T omitted for the moment): the box P at the end represents a repreparation of systems according to the estimated density matrix. The overall output will then be a quantum system, which can be directly compared with the inputs in statistical experiments. It is clear that the state cannot be determined exactly from a sample with finite N, but the determination becomes arbitrarily good in the limit $N \to \infty$. Optimal estimation observables are known in the case when the inputs are guaranteed to be pure [74], but in the case of general mixed states there are no clear-cut theorems yet, partly owing to the fact that it is less clear what "figure of merit" best describes the quality of such an estimator.

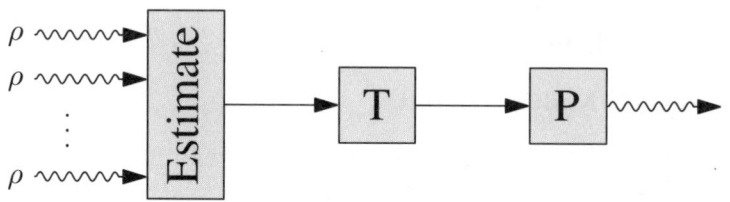

Fig. 2.5. Classical teleportation with multiple inputs, or state estimation

Given a good estimator we can, of course, proceed to good cloning by just repeating the repreparation P as often as desired. The surprise here [75] is that if only a fixed number M of outputs is required, it is possible to obtain better clones with devices that stay entirely in the quantum world than by going via classical estimation. Again, the problem of *optimal cloning* is fully understood for pure states [76], but work has only just begun to understand the mixed-state case.

Another operation which becomes accessible in this way is the *universal not* operation, assigning to each pure qubit state the unique pure state orthogonal to it. Like time reversal, this is just a special case of an antiunitary symmetry operation. In this case, a strategy using a classical estimation as an intermediate step can be shown to be optimal [77]. In this sense "universal not" is a harder task than "cloning".

More generally, we can look at schemes such as those in Fig. 2.5, where T represents any transformation of the density matrix data, whether or not this transformation corresponds to a physically realizable transformation of quantum states. A further interesting application is to the *purification* of states. In this problem it is assumed that the input states were once pure, but were later corrupted in some noisy environment (the same for all inputs). The task is to reconstruct the original pure states. Usually, the noise corresponds to an invertible linear transformation of the density matrices, but its inverse is not a possible operation, because it transforms some density matrices to operators with negative eigenvalues. So the reversal of noise is not possible with a one-shot device, but is easy to perform to high accuracy when many equally prepared inputs are available. In the simplest case of a so-called depolarizing channel, this problem is well understood [78]; it is also well understood in the version requiring many outputs, as in the optimal-cloning problem [79].

2.4.2 Quantum Cryptography

It may seem impossible to find applications of impossible machines. But that is not quite true: sometimes the impossibility of a certain task is precisely what is called for in an application. A case in point is cryptography: here one tries to make the deciphering of a code impossible. So if we could design a code whose breaking would require one of the machines described in the previous section, we could guarantee its security *with the certainty of natural law*. This is precisely what *quantum cryptography* sets out to do. Because only small quantum systems are involved it is one of the "easiest" applications of quantum information ideas, and was indeed the first to be realized experimentally. For a detailed description we refer to Chap. 3. Here we just describe in what sense it is the application of an impossible machine.

As always in cryptography, the basic situation is that two parties, Alice and Bob, say, want to communicate without giving an "evil eavesdropper", conventionally named Eve, a chance to listen in. What classical eavesdroppers do is to tap the transmission line, *make a copy* of what they hear for later analysis, and otherwise let the signal pass undisturbed to the legitimate receiver (Bob). But if the signal is quantum, the no-cloning theorem tells us that faithful copying is impossible. So either Eve's copy or Bob's copy is corrupted. In the first case Eve won't learn anything, and hence there was no eavesdropping anyway. In the second case Bob will know that something may have gone wrong, and will tell Alice that they must discard that part of the secret key which they were exchanging. Of course, intermediate situations are possible, and one has to show very carefully that there is an exact trade-off between the amount of information Eve can get and the amount of perturbation she must inflict on the channel.

2.4.3 Entanglement-Assisted Teleportation

This is arguably the first major discovery in the field of quantum information. The no-cloning and no-teleportation theorems, although they had not been formulated in such terms, would hardly have come as a surprise to people working on the foundations of quantum mechanics in the 1960s, say. But entanglement assistance was really an unexpected turn. It was first seen by Bennett et al. [52], who also coined the term "teleportation". It is gratifying to see, though it is hardly a surprise on the same scale, that this prediction of quantum mechanics has also been implemented experimentally. The experiments are another interesting story, which will no doubt be told much better

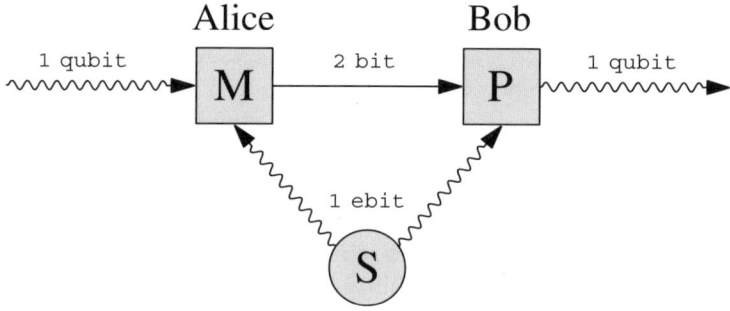

Fig. 2.6. Entanglement-assisted teleportation

in Chap. 3 by Weinfurter and Zeilinger, who represent one team in which a major breakthrough in this regard was achieved.

The teleportation scheme is shown in Fig. 2.6. What makes it so surprising is that it combines two machines whose impossibility was discussed in the previous section: omitting the distribution of the entangled state (the lower half of Fig. 2.6), we get the impossible process of classical teleportation. On the other hand, if we omit the classical channel, we get an attempt to transmit information by means of correlations alone, i.e. a version of Bell's Telephone. Since the time dimension is not represented in this diagram, let us consider the steps in the proper order. The first step is that Alice and Bob each receive one half of an entangled system. The source can be a third party or can be Bob's lab. The last choice is maybe best for illustrative purposes, because it makes clear that no information is flowing from Alice to Bob at this stage. Alice is next given the quantum system whose state (which is unknown to her) she is to teleport. Alice then makes a measurement on a system made by combining the input and her half of the entangled system. She sends the results via a classical channel to Bob, who uses them to adjust the settings on his device, which then performs some unitary transformation on his half of the entangled system. The resulting system is the output, and if everything is chosen in the right way, these output systems are indeed statistically indistinguishable from the outputs. To see just how the entangled state S, the measurement M and the repreparation P have to be chosen requires the mathematical framework of quantum theory. In the standard example one teleports a state of one qubit, using up one maximally entangled two-qubit system (in the current jargon, "1 ebit") and sending two classical bits from Alice to Bob. A general characterization of the teleportation schemes for qubits and higher-dimensional systems is given below in Sect. 2.6.6.

2.4.4 Superdense Coding

It is easy to see and in fact is a commonplace occurrence that classical information can be transmitted on quantum channels. For example, one bit of classical information can be coded in any two-level system, such as the polarization degree of freedom of a photon. It is not entirely trivial to prove, but hardly surprising that one cannot do better than "one bit per qubit". Can we beat this bound using the idea of entanglement assistance? It turns out that one can. In fact, one can double the amount of classical information carried by a quantum channel ("two bits per qubit"). Remarkably, the setups for doing this are closely related to teleportation schemes, and in the simplest cases Alice and Bob just have to swap their equipment for entanglement assisted teleportation. This is explained in detail in Sect. 2.6.6.

2.4.5 Quantum Computation

Again, we shall be very brief on this subject, although it is certainly central to the field. After all, it is partly the promise of a fantastic new class of computers which has boosted the interest in quantum information in recent years. But since in this book computation is covered in Chap. 4, we shall only make a few remarks connecting this subject to the theme of possible versus impossible machines.

So can quantum computers perform otherwise impossible tasks? Not really, because in principle we can solve the dynamical equations of quantum mechanics on a classical computer and simulate all the results. Hence classically unsolvable problems such as the halting problem for Turing machines and the word problem in group theory cannot be solved on quantum computers either. But this argument only shows the *possibility* of emulating all quantum computations on a classical computer, and omits the possibility that the efficiency of this procedure may be terrible. The great promise of quantum computation lies therefore in the reduction of running time, from exponential to polynomial time in the case of Shor's factorization algorithm [80]. This reduction is comparable to replacing the task of counting all the way up to a 137 digit number by just having to write it. No matter what the constants are in the growth laws for the computing time (and they will probably not be very favorable for the quantum contestant), the polynomial time is going to win if we are really interested in factoring very large numbers.

A word of caution is necessary here concerning the impossible/possible distinction. While it is true that no polynomial-time classical factoring algorithm is known, and this is what counts from a practical point of view, there is no *proof* that no such algorithm exists. This is a typical state of affairs in complexity theory, because the nonexistence of an algorithm is a statement about the rather unwieldy set of all Turing machine programs. A proof by inspecting all of them is obviously out, so it would have to be based on some principle of "conservation of difficulties", which rarely exists for real-life problems. One problem in which this is possible is that of identifying which (unique) element of a large list has a certain property ("needle in a haystack"). In this case the obvious strategy of inspecting every element in turn can be shown to be the optimal classical one, and has a running time proportional to the length N of the list. But Grover's quantum algorithm [58] performs the task in the order of $\sqrt{N}$ steps, an amazing gain even if it is not exponential. Hence there are problems for which quantum computers are provably faster than any classical computer.

So what makes the reduction of running time work? This is not so easy to answer, even after working through Shor's algorithm and verifying the claim of exponential speedup. Massive entanglement is used in the algorithm, so this is certainly one important element. Then there is a technique known as *quantum parallelism*, in which a quantum computation is run on a coherent superposition of all possible classical inputs, and, in a sense, all values of a function are computed simultaneously. A catchy paraphrase, attributed to D. Deutsch, is to call this a computation in the parallel worlds of the many-worlds interpretation.

But perhaps the best way to find out what powers quantum computation is to to turn it around and to really try the classical emulation. The practical difficulty which then becomes apparent immediately is that the Hilbert space dimensions grow extremely fast. For N qubits (two-level systems), one has to operate in a Hilbert space of 2^N dimensions. The corresponding space of density matrices has 2^{2N} dimensions. For classical bits one has instead a configuration space of 2^N discrete points, and the analogue of the density matrices, the probability densities, live in a merely 2^N-dimensional space. Brute force simulations of the whole system therefore tend to grind to a halt even on fairly small systems. Feynman was the first to turn this around: maybe only a quantum system can be used to simulate a quantum system, and maybe, while we are at it, we can go beyond simulation and do some interesting computations as well. So, putting it positively, in a quantum system we have exponentially more dimensions to work with: *there is lots of room in Hilbert space*. The added complexity of quantum versus classical correlations, i.e. the phenomenon of entanglement, is also a consequence of this.

But it is not so easy to use those extra dimensions. For example, for transmission of classical information an N-qubit system is no better than a classical N-bit system. Only the entanglement assistance of superdense coding brings out the additional dimensions. Similarly, quantum computers do not speed up every computation, but are good only at specific tasks where the extra dimensions can be brought into play.

2.4.6 Error Correction

Again, we shall only make a few remarks related to the possible/impossible theme, and refer the reader to Chap. 4 for a deeper discussion. First of all, error correction is absolutely crucial for the implementation of quantum computers. Very early in the development of the subject the suspicion was raised that exponential speedup was only possible if all component parts of the computer were realized with exponentially high (and hence practically unattainable) precision.

In a classical computer the solution to this problem is *digitization*: every bit is realized by a bistable circuit, and any deviation from the two wanted states is restored by the circuit at the expense of some energy and with some heat generation. This works separately for every bit, so in a sense every bit has its own heat bath. But this strategy will not work for quantum computers: to begin with, there is now a continuum of pure states which would have to be stabilized for every qubit, and, secondly, one heat bath per qubit would quickly destroy entanglement and hence make the quantum computation impossible. There are many indications that entanglement is indeed more easily destroyed by thermal noise and other sources of errors; this is summarily referred to as *decoherence*. For example, a Gaussian channel (this is a special type of infinite-dimensional channel) has infinite capacity for classical information, no matter how much noise we add. But its quantum capacity drops to zero if we add more classical noise than that specified by the Heisenberg uncertainty relations [81].

A standard technique for stabilizing classical information is *redundancy*: just send a classical bit three times, and decide at the end by majority vote which bit to take. It is easy to see that this reduces the probability of error from order ε to order ε^2. But quantum mechanically, this procedure is forbidden by the no-cloning theorem: we simply cannot make three copies to start the process.

Fortunately, quantum error correction is possible in spite of all these doubts [82]. Like classical error correction, it also works by distributing the quantum information over several parallel channels, but it does this in a much more subtle way than copying. Using five parallel channels, one can obtain a similar reduction of errors from order ε to order ε^2 [63]. Much more has been done, but many open questions remain, for which I refer once again to Chap. 4.

2.5 A Preview of the Quantum Theory of Information

Before we go on in the next section to turn some of the heuristic descriptions of the previous sections into rigorous mathematical statements, I shall try to give a flavor of the theory to be constructed, and of its motivations and current state of development.

Theoretical physics contributes to the field of quantum information processing in two distinct, though interrelated ways. One of these ways is the construction of *theoretical models* of the systems which are being set up experimentally as candidates for quantum devices. Of course, any such system will have very many degrees of freedom, of which only very few are singled out as the "qubits" on which the quantum computation is performed. Hence it is necessary to analyze to what degree and on what timescales it is justified to treat the qubit degrees of freedom separately, and with what errors the desired quantum operation can be realized in the given system. These questions are crucial for the realization of any quantum devices, and require specialized in-depth knowledge of the appropriate theory, e.g. quantum optics, solid-state theory or quantum chemistry (in the case of NMR quantum computing). However, these problems are not what we want to look at in this chapter. The other way in which theoretical physics contributes to the field of quantum information processing is in the form of another kind of theoretical work, which could be called the "abstract quantum theory of information". Recall the arguments in Sect. 2.2, where the possibility of translating between different carriers of (classical) information was taken as the justification for looking at an abstracted version, the classical theory of information, as founded by Shannon. While it is true that quantum information cannot be translated into this framework, and is hence a new kind of information, translation is often possible (at least in principle) between different carriers of quantum information. Therefore, we can make a similar abstraction in the quantum case. To this abstract theory all qubits are the same, whether they are realized as polarizations of photons, nuclear spins, excited states of ions in a trap, modes of a cavity electromagnetic field or whatever other realization may be feasible. A large amount of work is currently being devoted to this abstract branch of quantum information theory, so I shall list some of the reasons for this effort.

- Abstract quantum theoretical reasoning is how it all started. In the early papers of Feynman and Deutsch, and in the papers by Bennett and coworkers, it is the structure of quantum theory itself which opens up all these new possibilities. No hint from experiment and no particular theoretical difficulty in the description of concrete systems prompted this development. Since the technical realizations are lagging behind so much, the field will probably remain "theory driven" for some time to come.
- If we want to transfer ideas from the classical theory of information to the quantum theory, we shall always get abstract statements. This works quite well for importing good questions. Unfortunately, however, the answers are most of the time not transferred so easily.
- The reason for this difficulty with importing classical results is that some of the standard probabilistic techniques, such as conditioning, do not work in quantum theory, or work only sporadically. This is the same problem that the statistical mechanics of quantum many-particle systems faces in comparison with its classical sister. The cure can only be the development of new, genuinely quantum techniques. Preferably these should work in the widest (and hence most abstract) possible setting.

- One of the fascinating aspects of quantum information is that features of quantum mechanics which were formerly seen only as paradoxical or counterintuitive are now turned into an asset: these are precisely the features one is trying to utilize now. But this means that naive intuitive reasoning tends to lead to wrong results. Until we know much more about quantum information, we shall need rigorous guidance from a solid conceptual and mathematical foundation of the theory.
- When we take as a selling point for, say, quantum cryptography that secrets are protected "with the security of natural law", the argument is only as convincing as the *proof* that reduces this claim to first principles. Clearly this requires abstract reasoning, because it must be independent of the physical implementation of the device the eavesdropper uses. The argument must also be completely rigorous in the mathematical sense.
- Because it does not care about the physical realization of its "qubits", the abstract quantum theory of information is applicable to a wide range of seemingly very different systems. Consider, for example, some abstract quantum gate like the "controlled not" (C-NOT). From the abstract theory, we can hope to obtain relevant quality criteria, such as the minimal fidelity with which this device has to be implemented for some algorithm to work. So systems of quite different types can be checked according to the same set of criteria, and a direct competition becomes possible (and interesting) between different branches of experimental physics.

So what will be the basic concepts and features of the emerging quantum theory of information? The information-theoretical perspective typically generates questions like

How can a given task of quantum information processing be performed optimally with the given resources?

We have already seen a few typical tasks of quantum information processing in the previous section and, of course, there are more. Typical resources required for cryptography, quantum teleportation and dense coding are entangled states, quantum channels and classical channels. In error correction and computing tasks, the resources are the size of the quantum memory and the number of quantum operations. Hence all these notions take on a quantitative meaning.

For example, in entanglement-assisted teleportation the entangled pairs are used up (one maximally entangled qubit pair is needed for every qubit teleported). If we try to run this process with less than maximally entangled states, we may still ask how many pairs from a given preparation device are needed per qubit to teleport a message of many qubits, say, with an error less than ε. This quantity is clearly a measure of entanglement. But other tasks may lead to different quantitative measures of entanglement. Very often it is possible to find inequalities between different measures of entanglement, and establishing these inequalities is again a task of quantum information theory.

The direct definition of an entanglement measure based on teleportation, of the quantum information capacity of a channel and of many similar quantities requires an optimization with respect to all codings and decodings of asymptotically long quantum messages, which is extremely hard to evaluate. In the classical case, however, there is a simple formula for the capacity of a noisy channel, called Shannon's coding theorem, which allows us to compute the capacity directly from the transition probabilities of a channel. Finding quantum analogues of the coding theorem (and similar formulas for entanglement resources) is still one of the great challenges in quantum information theory.

2.6 Elements of Quantum Information Theory

It is probably too early to write a definitive account of quantum information theory – there are simply too many open questions. But the basic concepts are clear enough, and it will be the task of the remainder of this chapter to explain them, and use the precise definitions to state some of the interesting open questions in the field. In the limited space available this cannot be done in textbook style, with many examples and full proofs of all the things used on the way (or even full references of them). So I shall try to emphasize the main lines and to set up the basic definitions using as few primitive concepts as possible. For example, the capacities of a channel for either classical or quantum information will be defined on exactly the same pattern. This will make it easier to establish the relations between these concepts.

The following sections begin with material which every physicist knows from quantum mechanics courses, although maybe not in this form. We need to go over this material, though, in order to establish the notation.

2.6.1 Systems and States

The systems occurring in the theory can be either quantum or classical, or can be hybrids composed of a classical and a quantum part. Therefore, we need a mathematical framework covering all these cases. A good choice is to characterize each type of system by its *algebra of observables*. In this chapter all algebras of observables will be taken to be *finite-dimensional* for simplicity. Extensions to infinite dimensionality are mostly straightforward, though, and in fact a strength of the algebraic approach to quantum theory is that it deals not just with infinite-dimensional algebras, but also with systems of infinitely many degrees of freedom, as in quantum field theory [83,84] and statistical mechanics [85].

The first main type of system consists of purely *classical systems*, whose algebra of observables is commutative, and can hence be considered as a space of complex-valued functions on a set X. Our assumption of finiteness requires that X is a finite set, and the algebra of observables $\mathcal{A}$ will be $\mathcal{C}(X)$, the space of all functions $f : X \to \mathbb{C}$. A single *classical bit* corresponds to the choice $X = \{0, 1\}$. On the other hand, a purely *quantum system* is determined by the choice $\mathcal{A} = \mathcal{B}(\mathcal{H})$, the algebra of all bounded linear operators on the Hilbert space $\mathcal{H}$. The finiteness assumption requires that $\mathcal{H}$ has a finite dimension d, so $\mathcal{A}$ is just the space $\mathcal{M}_d$ of complex $d \times d$ matrices. A *qubit* is given by $\mathcal{A} = \mathcal{M}_2$.

The basic statistical interpretation of the algebra of observables is the same in the quantum and classical cases, and hinges on the cone of positive elements in the algebra. Here Y is called *positive* (in symbols, $Y \geq 0$) if it can be written in the form $Y = X^*X$. Then $Y \in \mathcal{M}_d$ is positive exactly if it is given by a positive semidefinite matrix, and $f \in \mathcal{C}(X)$ is positive iff $f(x) \geq 0$ for all x. In any algebra of observables $\mathcal{A}$, we shall denote by $\mathbb{1} \in \mathcal{A}$ the identity element.

A *state* Φ on $\mathcal{A}$ is a positive normalized linear functional on $\mathcal{A}$. That is, $\Phi : \mathcal{A} \to \mathbb{C}$ is linear, with $\Phi(X^*X) \geq 0$ and $\Phi(\mathbb{1}) = 1$. Each state describes a way of preparing systems, in all the details that are relevant to subsequent statistical measurements on the systems. The measurements are described by assigning to each outcome from a device an *effect* $F \in \mathcal{A}$, i.e. an element with $0 \leq F \leq \mathbb{1}$. The prediction of the theory for the probability of that outcome, measured on systems prepared according to the state ρ, is then $\rho(F)$.

For explicit computations we shall often need to expand states and elements of $\mathcal{A}$ in a basis. The standard basis in $\mathcal{C}(X)$ consists of the functions $e_x, x \in X$, such that $e_x(y) = 1$ for $x = y$ and zero otherwise. Similarly, if $\phi_\mu \in \mathcal{H}$ is an orthonormal basis of the Hilbert space of a quantum system, we denote by $e_{\mu\nu} = |e_\mu\rangle\langle e_\nu| \in \mathcal{B}(\mathcal{H})$ the corresponding "matrix units". Then a state p on the classical algebra $\mathcal{C}(X)$ is characterized by the numbers $p_x \equiv p(e_x)$, which form a probability distribution on X, i.e. $p(x) \geq 0$ and $\sum_x p(x) = 1$. Similarly, a quantum state ρ on $\mathcal{B}(\mathcal{H})$ is given by the numbers $\rho_{\mu\nu} \equiv \rho(e_{\nu\mu})$, which form the so-called *density matrix*. If we interpret them as the expansion coefficients of an operator $\widehat{\rho} = \sum_{\mu\nu} \rho_{\mu\nu} e_{\mu\nu}$, the *density operator* of ρ, we can also write $\rho(A) = \mathrm{tr}(\widehat{\rho}A)$.

A state is called *pure* if it is extremal in the convex set of all states, i.e. if it cannot be written as a convex combination $\lambda \rho' + (1 - \lambda)\rho''$ of other states. These are the states which contain as little randomness as possible. In the classical case, the only pure states are those concentrated on a single point $z \in X$, i.e. $p_z = 1$, or $p(f) = f(z)$. The pure states in the quantum case are determined by "wave vectors" $\psi \in \mathcal{H}$ such that $\rho(A) = \langle \psi, A\psi \rangle$, and $\widehat{\rho} = |\psi\rangle\langle\psi|$. Thus, in the simplest case of a classical bit, there are just two extreme points, whereas in the case of a qubit the extreme points form a sphere in three dimensions and are given by the expectations of the three Pauli matrices:

$$\widehat{\rho} = \frac{1}{2}\begin{pmatrix} 1 + x_3 & x_1 - ix_2 \\ x_1 + ix_2 & 1 - x_3 \end{pmatrix} = \frac{1}{2}(\mathbb{1} + \boldsymbol{\sigma} \cdot \boldsymbol{x}),$$
$$x_k = \rho(\sigma_k). \tag{2.5}$$

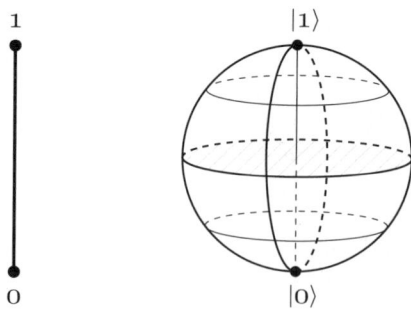

Fig. 2.7. State spaces as convex sets: *left*, one classical bit; *right*, one quantum bit (qubit)

Then positivity requires $|\boldsymbol{x}|^2 \leq 1$, with equality when ρ is pure. This is shown in Fig. 2.7.

Thus, in addition to the north pole $|1\rangle$ and the south pole $|0\rangle$, which roughly correspond to the extremal states of the classical bit, we have their coherent superpositions corresponding to the wave vectors $\alpha|1\rangle + \beta|0\rangle$, where $\alpha, \beta \in \mathbb{C}$, and $|\alpha|^2 + |\beta|^2 = 1$. This additional freedom becomes even more dramatic in higher-dimensional systems, and is crucial for the possibility of entanglement.

Entanglement is a property of states of composite systems, so we must introduce the notion of *composition* of systems. We shall define this in a way which applies to classical and quantum systems alike. If $\mathcal{A}$ and $\mathcal{B}$ are the algebras of observables of the subsystems, the algebra of observables of the composition is defined to be the tensor product $\mathcal{A} \otimes \mathcal{B}$. In the finite-dimensional case, which is our main concern, this is defined as the space of linear combinations of elements that can be written as $A \otimes B$ with $A \in \mathcal{A}$ and $B \in \mathcal{B}$, such that $A \otimes B$ is linear in A and linear in B. The algebraic operations are defined by $(A \otimes B)^* = A^* \otimes B^*$, and $(A_1 \otimes B_1)(A_2 \otimes B_2) = (A_1 A_2) \otimes (B_1 B_2)$. Thus $\mathbb{1} = \mathbb{1}_\mathcal{A} \otimes \mathbb{1}_\mathcal{B}$. Since positivity is defined in terms of a star operation (adjoint) and a product, these definitions also determine the states and effects of the composite system.

Let us explore how this unifies the more common definitions in the classical and quantum cases. For two classical factors $\mathcal{C}(X) \otimes \mathcal{C}(Y)$, a basis is formed by the elements $e_x \otimes e_y$, so the general element can be expanded as

$$f = \sum_{x,y} f(x,y) e_x \otimes e_y \,,$$

and each element can be identified with a function on the Cartesian product $X \times Y$. Hence $\mathcal{C}(X) \otimes \mathcal{C}(Y) \cong \mathcal{C}(X \times Y)$. Similarly, in the purely quantum case, we can expand in matrix units and obtain quantities with four indices: $(A \otimes B)_{\mu\nu,\mu'\nu'} = A_{\mu\mu'} B_{\nu\nu'}$. In a basis-free way, i.e. when A, B are considered as operators on Hilbert spaces $\mathcal{H}_A, \mathcal{H}_B$, this is defined by the equation

$$(A \otimes B)(\phi \otimes \psi) = (A\phi) \otimes (B\psi),$$

where $\phi \in \mathcal{H}_A$ and $\psi \in \mathcal{H}_B$, and the tensor product of the Hilbert spaces is formed in the usual way. Hence $\mathcal{B}(\mathcal{H}_A) \otimes \mathcal{B}(\mathcal{H}_B) \cong \mathcal{B}(\mathcal{H}_A \otimes \mathcal{H}_B)$.

But the definition of a composition by a tensor product of algebras of observables also determines how a quantum–classical *hybrid* must be described. Such systems occur frequently in quantum information theory, whenever a combination of classical and quantum information is given. We shall approach hybrids in two equivalent ways, which are also useful more generally. Suppose we know only that the first subsystem is classical and make no assumptions about the nature of the second, i.e. we want to characterize tensor products of the form $\mathcal{C}(X) \otimes \mathcal{B}$. Then every element can be expanded in the form $B = \sum_x e_x \otimes B_x$, where now $B_x \in \mathcal{B}$. Clearly, the elements B_x determine B, and hence we can identify the tensor product with the space (sometimes denoted by $\mathcal{C}(X; \mathcal{B})$) of $\mathcal{B}$-valued functions on X with pointwise algebraic operations. Similarly, suppose that we know only that $\mathcal{B} = \mathcal{M}_d$ is the algebra of $d \times d$ matrices. Then, expanding in matrix units, we find that $A = \sum_{\mu\nu} A_{\mu\nu} \otimes e_{\mu\nu}$ with $A_{\mu\nu} \in \mathcal{A}$. That is, we can identify $\mathcal{A} \otimes \mathcal{M}_d$ with the space (sometimes denoted by $\mathcal{M}_d(\mathcal{A})$) of $d \times d$ matrices with entries from $\mathcal{A}$. By using the relation $e_{\mu\nu} e_{\alpha\beta} = \delta_{\nu\alpha} e_{\mu\beta}$, one can readily verify that the product in $\mathcal{A} \otimes \mathcal{M}_d$ indeed corresponds to the usual matrix multiplication in $\mathcal{M}_d(\mathcal{A})$, with due care given to the order of factors in products with elements from $\mathcal{A}$, if $\mathcal{A}$ happens to be noncommutative. The adjoint is given by $(A^*)_{\mu\nu} = (A_{\nu\mu})^*$. Hence a hybrid algebra $\mathcal{C}(X) \otimes \mathcal{M}_d$ can be viewed either as the algebra of $\mathcal{C}(X)$-valued $d \times d$ matrices or as the space of $\mathcal{M}_d$-valued functions on X.

The physical interpretation of a composite system $\mathcal{A} \otimes \mathcal{B}$ in terms of states and effects is straightforward. When $F \in \mathcal{A}$ and $G \in \mathcal{B}$ are effects, so is $F \otimes G$, and this is interpreted as the joint measurement of F on the first subsystem and of G on the second subsystem, where the "yes" outcome is taken as "both effects give yes". In particular, $F \otimes \mathbb{1}_\mathcal{B}$ corresponds to measuring F on the first system, completely ignoring the second. Thus, for any state ρ on $\mathcal{A} \otimes \mathcal{B}$ we define the *restriction* $\rho_\mathcal{A}$ of ρ to $\mathcal{A}$ by $\rho_\mathcal{A}(A) = \rho(A \otimes \mathbb{1}_\mathcal{B})$. In the classical case, the probability density for $\rho_\mathcal{A}$ is obtained by integrating out the $\mathcal{B}$ variables. In the quantum case, it corresponds to the partial trace of density matrices with respect to $\mathcal{H}_B$. In general, it is not possible to reconstruct the state ρ from the restrictions $\rho_\mathcal{A}$ and $\rho_\mathcal{B}$, which is another way of saying that ρ also describes correlations between the systems. However, given $\rho_\mathcal{A}$ and $\rho_\mathcal{B}$, there is always a state with these restrictions, namely the tensor product $\rho_\mathcal{A} \otimes \rho_\mathcal{B}$, which corresponds to an independent preparation of the subsystems.

A fundamental difference between quantum and classical correlations lies in the nature of pure states of composite systems. Classically the situation is easy: a pure state of the composite system $\mathcal{C}(X) \otimes \mathcal{C}(Y) \cong \mathcal{C}(X \times Y)$ is just a point $(x, y) \in X \times Y$. Obviously, the restrictions of this state are the pure states concentrated on x and y, respectively. More generally, whenever one of the algebras in $\mathcal{A} \otimes \mathcal{B}$ is commutative, every pure state will restrict to pure states on the subsystems. Not so in the purely quantum case. Here the pure states are given by unit vectors Φ in the tensor product $\mathcal{H}_A \otimes \mathcal{H}_B$, and unless Φ happens to be of the special form $\phi_A \otimes \phi_B$ (and not a linear combination of such vectors), the state will not be a product, and the restrictions will not be pure. The following standard form of vectors in a tensor product, known as the *Schmidt decomposition*, is used in entanglement theory every day, and twice on Sundays.

Lemma 2.1. *(1) ("Schmidt decomposition") Let $\Phi \in \mathcal{H}_A \otimes \mathcal{H}_B$ be a unit vector, and let $\widehat{\rho}_A$ denote the density operator of its restriction to the first factor. Then if $\widehat{\rho}_A = \sum_\mu \lambda_\mu |e_\mu\rangle\langle e_\mu|$ (with $\lambda_\mu > 0$) is the spectral resolution, we can find an orthonormal system $e'_\mu \in \mathcal{H}_B$ such that*

$$\Phi = \sum_\mu \sqrt{\lambda_\mu}\, e_\mu \otimes e'_\mu .$$

(2) ("Purification") An arbitrary quantum state ρ on $\mathcal{H}$ can be extended to a pure state on a larger system with a Hilbert space $\mathcal{H} \otimes \mathcal{H}_B$. Moreover, the restricted density matrix $\widehat{\rho}_B$ can be chosen to have no zero eigenvalues, and with this additional condition the space $\mathcal{H}_B$ and the extended pure state are unique up to a unitary transformation.

Proof. (1) We may expand Φ as $\Phi = \sum_\mu e_\mu \otimes \psi_\mu$, with suitable vectors $\Psi_\mu \in \mathcal{H}_B$. The reduced density matrix is determined by

$$\mathrm{tr}(\widehat{\rho}_A F) = \langle \Phi, (A \otimes \mathbb{1})\Phi \rangle = \sum_{\mu\nu} \langle e_\mu, A e_\nu \rangle \langle \psi_\mu, \psi_\nu \rangle = \sum_\mu \lambda_\mu \langle e_\mu, A e_\mu \rangle .$$

Since A is arbitrary (e.g. $A = |e_\alpha\rangle\langle e_\beta|$), we may compare coefficients, and obtain $\langle \psi_\mu, \psi_\nu \rangle = \lambda_\mu \delta_{\mu\nu}$. Hence $e'_\mu = \lambda^{-1/2} \psi_\mu$ is the desired orthonormal system.

(2) The existence of the purification is evident if one defines Φ as above, with the orthonormal system e'_μ chosen in an arbitrary way. Then $\widehat{\rho}_B = \sum_\mu \lambda_\mu |e'_\mu\rangle\langle e'_\mu|$, and the above computation shows that choosing the basis e_μ is the only freedom in this construction. But any two bases are linked by a unitary transformation. □

A nonproduct pure state is a basic example of an entangled state in the sense of the following definition:

Definition 2.1. *A state ρ on $\mathcal{A} \otimes \mathcal{B}$ is called* separable *(or "classically correlated") if it can be written as*

$$\rho = \sum_\mu \lambda_\mu\, \rho^A_\mu \otimes \rho^B_\mu ,$$

with states ρ_μ^A and ρ_μ^B on $\mathcal{A}$ and $\mathcal{B}$, respectively, and weights $\lambda_\mu > 0$. Otherwise, ρ is called entangled.

Thus a classically correlated state may well contain nontrivial correlations. In fact, if either $\mathcal{A}$ or $\mathcal{B}$ is classical, *every* state is classically correlated. What the definition expresses is only that we may generate these correlations by a purely classical mechanism. We can use a classical random generator, which produces the result "μ" with probability λ_μ, together with two preparing devices operating independently but receiving instructions from the random generator: ρ_μ^A is the state produced by the $\mathcal{A}$ device if it receives the input "μ" from the random generator, and similarly for $\mathcal{B}$. Then the overall state prepared by this setup is ρ, and clearly the source of all correlations in this state lies in the classical random generator.

For an extensive treatment of these concepts the reader is now referred to Chap. 5. We shall turn instead to the second fundamental type of objects in quantum information theory, the channels.

2.6.2 Channels

Any processing step of quantum information is represented by a "channel". This covers a great variety of operations, from preparation to time evolution, measurement, and measurement with general state changes. Both the input and the output of a channel may be an arbitrary combination of classical and quantum information. The combination of different kinds of inputs or outputs causes no special problems of formulation: it simply means that the algebras of observables of the input and output systems of a channel must be chosen as suitable tensor products.

The basic idea of the mathematical description of a channel is to characterize the transformation T in terms of the way it modifies subsequent measurements. Suppose the channel converts systems with algebra $\mathcal{A}$ into systems with algebra $\mathcal{B}$. Then, by applying first the channel, and then a yes/no measurement F on the $\mathcal{B}$-type output system, we have effectively measured an effect on the $\mathcal{A}$-type system, which will be denoted by $T(F)$. Hence a channel is completely specified by a map $T : \mathcal{B} \to \mathcal{A}$, and we shall say, for simplicity, that this map *is* the channel. There is, of course an alternative way of viewing a channel, namely as a map taking input states to output states, i.e. states on $\mathcal{A}$ into states on $\mathcal{B}$, which we we shall denote by T_*. We shall say that T describes the channel in the *Heisenberg picture*, whereas T_* describes the same channel in the *Schrödinger picture*. The descriptions are connected by the equation

$$[T_*(\rho)](F) = \rho[T(F)] \tag{2.6}$$

where ρ is an arbitrary state on $\mathcal{A}$, and $F \in \mathcal{B}$ is also arbitrary. The notation on the left-hand side is sometimes a little clumsy; therefore we shall often write $T_*(\rho) = \rho \circ T$, where "∘" denotes a composition of maps, in this case from $\mathcal{B}$ to $\mathcal{A}$ to $\mathbb{C}$. A *composition of channels* will then also be written in the form $S \circ T$. This has the advantage that things are written from left to right in the order in which they happen: first the preparation, then some channels, and finally the yes/no measurement F. As a further simplification, we shall often follow the convention of dropping the parentheses of the arguments of linear operators (e.g. $T(A) \equiv TA$) and dropping the ∘ symbols, but reintroducing any of these elements for punctuation whenever they help to make expressions unambiguous or just more readable.

For many questions in quantum information theory it is crucial to have a precise notion of the set of possible channels between two types of systems: clearly, the distinction between "possible" and "impossible" in Sect. 2.3 is of this kind, but the search for the "optimal device" performing a certain task is also of this kind. There are two different approaches for defining the set of maps $T : \mathcal{B} \to \mathcal{A}$ which should qualify as channels, and luckily they agree. The first approach is *axiomatic*: one just lists the properties of T which are forced on us by the statistical interpretation of the theory. The second approach is *constructive*: one lists operations which can actually be performed according to the conventional wisdom of quantum mechanics and defines the admissible channels as those which can be assembled from these building blocks. The equivalence between these approaches is one of the fundamental theorems in this field, and is known as the Stinespring dilation theorem. We shall state this theorem after describing both approaches and giving a formal definition of "channels".

Note that the left-hand side of (2.6) is linear in F, which reflects the fact that a mixture of effects ("use effect F_1 in 42% of the cases and F_2 in the remaining cases") directly becomes a mixture of the corresponding probabilities. Therefore, the right-hand side also has to be linear in F, i.e. $T : \mathcal{B} \to \mathcal{A}$ must be a linear operator, from the statistical interpretation of the theory. Obviously, T also has to take positive operators F into a positive $T(F)$ ("T is positive"), and the trivial measurement has to remain trivial: $T\mathbb{1}_\mathcal{B} = \mathbb{1}_\mathcal{A}$ ("T is unit preserving, or unital"). This is equivalent to T_* being likewise a positive linear operator, with the normalization condition $\operatorname{tr} T_*(\rho) = \operatorname{tr} \rho$. Finally, we would like to have an operation of "running two channels in parallel", i.e. we would like to define $T \otimes S : \mathcal{A}_1 \otimes \mathcal{B}_1 \to \mathcal{A}_2 \otimes \mathcal{B}_2$ for arbitrary channels $T : \mathcal{A}_1 \to \mathcal{A}_2$ and $S : \mathcal{B}_1 \to \mathcal{B}_2$. Since the identity $\mathbb{I}_n$ on an n-level quantum system $\mathcal{M}_n$ is one of the channels we want to consider, we must demand that $T \otimes \mathbb{I}_n$ also takes positive elements to positive elements. This "complete positivity" of T is a nontrivial requirement for maps between quantum systems. If $\mathcal{A}$ or $\mathcal{B}$ is classical, any positive linear map from $\mathcal{A}$ to $\mathcal{B}$ is automatically completely positive. For arbitrary completely positive maps, the product $T \otimes S$ is defined and is again completely positive, so just requiring the ability to form a tensor product with the "innocent bystander" $\mathbb{I}_n$ suffices to make all parallel channels well defined.

Definition 2.2. *A channel* converting systems with an algebra of observables $\mathcal{A}$ to systems with an algebra of observables $\mathcal{B}$ *is a completely positive, unit-preserving linear operator* $T: \mathcal{B} \to \mathcal{A}$.

In the "constructive" approach one allows only maps which can be built from the basic operations of (1) tensoring with a second system in a specified state, (2) unitary transformation and (3) reduction to a subsystem. Let us describe these and some other basic channels more formally, if only to show the richness of this concept. We leave the verification of the channel properties, including complete positivity, to the reader.

- *Expansion.* This expands system $\mathcal{A}$ by system $\mathcal{B}$ in the state ρ', say. Thus $T_*(\rho) = \rho \otimes \rho'$, or, from (2.6), $T: \mathcal{A} \otimes \mathcal{B} \to \mathcal{A}$ with $T(A \otimes B) = \rho'(B)A$.

- *Restriction.* In the Heisenberg picture, the operation of discarding system $\mathcal{B}$ from the composite system $\mathcal{A} \otimes \mathcal{B}$ is $T: \mathcal{A} \to \mathcal{A} \otimes \mathcal{B}$, with $T(A) = A \otimes \mathbb{1}_\mathcal{B}$. As noted before, this corresponds to taking partial traces if $\mathcal{B}$ is quantum, and to an integration over Y if $\mathcal{B} = \mathcal{C}(Y)$ is classical.

- *Symmetry.* By definition, the symmetries of a quantum system with an algebra of observables $\mathcal{A}$ are the invertible channels, i.e. channels $T: \mathcal{A} \to \mathcal{A}$ such that there is a channel S with $ST = TS = \mathbb{I}_\mathcal{A}$. It turns out that these are precisely the automorphisms of $\mathcal{A}$, i.e. invertible linear maps $T: \mathcal{A} \to \mathcal{A}$ such that $T(AB) = T(A)T(B)$, and $T(A^*) = T(A)^*$. For a pure quantum system the symmetries are precisely the unitarily implemented maps, i.e. the maps of the form $T(A) = UAU^*$, where U is a unitary element of $\mathcal{A}$. To readers familiar with Wigner's theorem (e.g. corollary 3.3 in [86]), another class of maps is conspicuously absent here, namely positive maps of the form $T(A) = \Theta A^* \Theta^*$ where Θ is *antiunitary*. It is well known that owing to the positivity of energy, a *time-reversal symmetry* can be implemented only by such an antiunitary transformation. But since such symmetries are not *completely* positive, they can only be global symmetries, and can never occur as symmetries affecting only a subsystem of the world.

- *Observable.* A measurement is simply a channel with a classical output algebra, say $\mathcal{B} = \mathcal{C}(X)$. Obviously, $T: \mathcal{B} \to \mathcal{A}$ is uniquely determined by the collection of operators $F_x = T(e_x)$ via $Tf = \sum_x f(x) F_x$. The channel property of T is equivalent to

$$F_x \in \mathcal{A}, \quad F_x \geq 0, \quad \sum_x F_x = \mathbb{1}_\mathcal{A}.$$

Either the "resolution of the identity" $\{F_x\}$ or the channel T will be called an observable here. This differs in two ways from the usual textbook definitions of this term: firstly, the outputs $x \in X$ need not be real numbers, and secondly, the operators F_x, whose expectations are the probabilities for obtaining the output x, need not be projection operators. This is sometimes expressed by calling T a generalized observable or a POVM (positive operator-valued measure). This is to distinguish them from the old-style "nongeneralized" observables, which are called PVMs (projection-valued measures) because $F_x^2 = F_x$.

- *Separable Channel.* A classical teleportation scheme is a composition of an observable and a preparation depending on a classical input, i.e. it is of the form

$$T(A) = \sum_x F_x \otimes \rho_x(A) , \qquad (2.7)$$

where the F_x form an observable, and ρ_x is the reconstructed state when the measurement result is x. Equivalently, we can say that $T = RS$, where 'input of S' = 'output of R' is a classical system with an algebra of observables $\mathcal{C}(X)$. The impossibility of classical teleportation, in this language, is the statement that no separable channel can be equal to the identity.

- *Instrument.* An observable describes only the statistics of the measuring results, and contains no information about the state of the system after the measurement. If we want such a more detailed description, we have to count the quantum system after the measurement as one of the outputs. Thus we obtain a composite output algebra $\mathcal{C}(X) \otimes \mathcal{B}$, where X is the set of classical outcomes of the measurement and $\mathcal{B}$ describes the output systems, which can be of a different type in general from the input systems with an algebra of observables $\mathcal{A}$. The term "instrument" for such devices was coined by Davies [86]. As in the case of observables, it is convenient to expand in the basis $\{e_x\}$ of the classical algebra. Thus $T : \mathcal{C}(X) \otimes \mathcal{B} \to \mathcal{A}$ can be considered as a collection of maps $T_x : \mathcal{B} \to \mathcal{A}$, such that $T(f \otimes B) = \sum_x f(x) T_x(B)$. The conditions on T_x are

$$T_x : \mathcal{B} \to \mathcal{A} \quad \text{completely positive, and} \quad \sum_x T_x(\mathbb{1}) = \mathbb{1} .$$

Note that an instrument has two kinds of "marginals": we can ignore the $\mathcal{B}$ output, which leads to the observable $F_x = T_x(\mathbb{1}_\mathcal{B})$, or we can ignore the measurement results, which gives the overall state change $\bar{T} = \sum_x T_x : \mathcal{B} \to \mathcal{A}$.

- *Von Neumann Measurement.* A von Neumann measurement is a special case of an instrument, associated with a family of orthogonal projections, i.e. $p_x \in \mathcal{A}$ with $p_x^* p_y = \delta_{xy} p_x$ and $\sum_x p_x = \mathbb{1}$. These define an instrument $T : \mathcal{C}(X) \otimes \mathcal{A} \to \mathcal{A}$ via $T_x(A) = p_x A p_x$. What von Neumann actually proposed [87] was the version of this with one-dimensional projections p_x, so the general case is sometimes called an incomplete von Neumann measurement or a Lüders measurement. The characteristic property of such measurements is their *repeatability*: since $T_x T_y = 0$ for $x \neq y$, repeating the measurement a second time (or any number of times) will always give the same output. For this reason the "projection postulate", which demanded that any decent measurement should be of this form, dominated the theory of quantum measurement processes for a long time.

- *Classical Input*. Classical information may occur in the input of a device just as well as in the output. Again this leads to a family of maps $T_x : \mathcal{B} \to \mathcal{A}$ such that $T : \mathcal{B} \to \mathcal{C}(X) \otimes \mathcal{A}$, with $T(B) = \sum_x e_x \otimes T_x(B)$. The conditions on $\{T_x\}$ are

$$T_x : \mathcal{B} \to \mathcal{A} \quad \text{completely positive, and} \quad T_x(\mathbb{1}) = \mathbb{1} \,.$$

Note that this looks very similar to the conditions for an instrument, but the normalization is different. An interesting special case is a "preparator", for which $\mathcal{A} = \mathbb{C}$ is trivial. This prepares $\mathcal{B}$ states that depend in an arbitrary way on the classical input x.

- *Kraus Form*. Consider quantum systems with Hilbert spaces $\mathcal{H}_A$ and $\mathcal{H}_B$, and let $K : \mathcal{H}_A \to \mathcal{H}_B$ be a bounded operator. Then the map $T_K(B) = K^*BK$ from $\mathcal{B}(\mathcal{H}_B)$ to $\mathcal{B}(\mathcal{H}_A)$ is positive. Moreover, $T_K \otimes \mathbb{I}_n$ can be written in the same form, with K replaced by $K \otimes \mathbb{I}_n$. Hence T_K is completely positive. It follows that maps of the form

$$T(B) = \sum_x K_x^* B K_x \,, \quad \text{where} \quad \sum_x K_x^* K_x = \mathbb{1} \,, \tag{2.8}$$

are channels. It is be a consequence of the Stinespring theorem that *any* channel $\mathcal{B}(\mathcal{H}_B)$ to $\mathcal{B}(\mathcal{H}_A)$ can be written in this form, which we call the Kraus form, following current usage. This refers to the book [88], which is a still recommended early account of the notion of complete positivity in physics.

- *Ancilla Form*. As stated above, every channel, defined abstractly as a completely positive normalized map, can be constructed in terms of simpler ones. A frequently used decomposition is shown in Fig. 2.8. The input system is coupled to an auxiliary system A, conventionally called the "ancilla" ("maidservant"). Then a unitary transformation is carried out, e.g. by letting the system evolve according to a tailor-made interaction Hamiltonian, and finally the ancilla (or, more generally, a suitable subsystem) is discarded.

The claim that *every* channel can be represented in the last two forms is a direct consequence of the fundamental structural theorem for completely

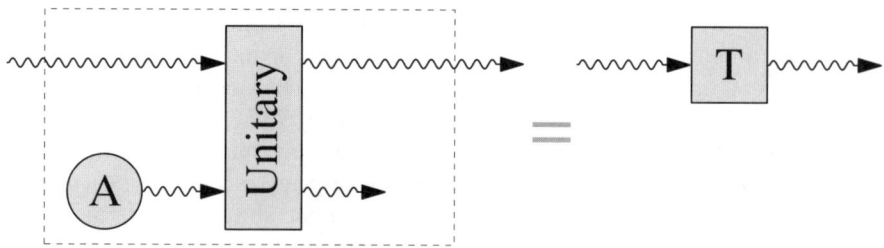

Fig. 2.8. Representation of an arbitrary channel as a unitary transformation on a system extended by an ancilla

positive maps, due to Stinespring [89]. We state it here in a version adapted to pure quantum systems, containing no classical components.

Theorem 2.1. (Stinespring Theorem). *Let $T : \mathcal{M}_n \to \mathcal{M}_m$ be a completely positive linear map. Then there is a number ℓ and an operator $V : \mathbb{C}^m \to \mathbb{C}^n \otimes \mathbb{C}^\ell$ such that*

$$T(X) = V^*(X \otimes \mathbb{1}_\ell)V , \qquad (2.9)$$

and the vectors of the form $(X \otimes \mathbb{1}_\ell)V\phi$, where $X \in \mathcal{M}_n$ and $\phi \in \mathbb{C}^m$, span $\mathbb{C}^n \otimes \mathbb{C}^\ell$. This decomposition is unique up to a unitary transformation of $\mathbb{C}^\ell$.

The ancilla form of a channel T is obtained by tensoring the Hilbert spaces $\mathbb{C}^m$ and $\mathbb{C}^n \otimes \mathbb{C}^\ell$ with suitable tensor factors $\mathbb{C}^a$ and $\mathbb{C}^b$, so that $ma = n\ell b$. One picks pure states in $\psi_a \in \mathbb{C}^a$ and $\psi_b \in \mathbb{C}^b$ and looks for a unitary extension of the map $\widetilde{V}\phi \otimes \psi_a = (V\phi) \otimes \psi_b$. There are many ways to do this, and this is a weakness of the ancilla approach in practical computations: one is always forced to specify an initial state ψ_a of the ancilla and many matrix elements of the unitary interaction, which in the end drop out of all results. As the uniqueness clause in the Stinespring theorem shows, it is the isometry V which neatly captures the relevant part of the ancilla picture.

In order to obtain the Kraus form of a general positive map T from its Stinespring representation, we choose vectors $\phi_x \in \mathbb{C}^\ell$ such that

$$\sum_x |\chi_x\rangle\langle\chi_x| = \mathbb{1} , \qquad (2.10)$$

and define Kraus operators K_x for T by $\langle\phi, K_x\psi\rangle = \langle\phi \otimes \chi_x, V\psi\rangle$ (we leave the straightforward verification of (2.8) to the reader). Of course, we can take the χ_x as an orthonormal basis of $\mathbb{C}^\ell$, but overcomplete systems of vectors do just as well.

It turns out that *all* Kraus decompositions of a given completely positive operator are obtained in the way just described. This follows from the following theorem, which solves the more general problem of finding all decompositions of a given completely positive operator into completely positive summands. In terms of channels, this problem has the following interpretation: for an instrument $\{T_x\}$, the sum $\bar{T} = \sum_x T_x$ describes the overall state change, when the measurement results are ignored. So the reverse problem is to find all measurements which are consistent with a given overall state change (perturbation) of the system, or, in physical terms, all *delayed-choice measurements* consistent with a given interaction between the system and its environment. By analogy with results for states on abelian algebras (probability measures) and states on C* algebras, we call the following theorem the Radon–Nikodym theorem. For a proof see [90].

Theorem 2.2. (Radon–Nikodym Theorem). *Let $T_x : \mathcal{M}_n \to \mathcal{M}_m$, $x \in X$ be a family of completely positive maps, and let $V : \mathbb{C}^m \to \mathbb{C}^n \otimes \mathbb{C}^\ell$ be the Stinespring operator of $\bar{T} = \sum_x T_x$. Then there are uniquely determined positive operators $F_x \in \mathcal{M}_\ell$ with $\sum_x F_x = \mathbb{I}$ such that*

$$T_x(X) = V^*(X \otimes F_x)V.$$

A simple but important special case is the case $\ell = 1$: then, since $\mathbb{C}^n \otimes \mathbb{C} \equiv \mathbb{C}^n$, we can just omit the tensor factor $\mathbb{C}^\ell$. The Stinespring form is then exactly that of a single term in the Kraus form with Kraus operator $K = V$. The Radon–Nikodym part of the theorem then says that the only decompositions of $\bar{T}$ into completely positive summands are decompositions into positive multiples of $\bar{T}$. Such maps are called "pure". Since the identity and, more generally, symmetries are of this type, we obtain the following corollary:

Corollary 2.1. (*"No information without perturbation"*). *Let $T : \mathcal{C}(X) \otimes \mathcal{M}_n \to \mathcal{M}_n$ be an instrument with a unitary global state change $\bar{T}(A) = T(1 \otimes A) = U^*AU$. Then there is a probability distribution p_x such that $T_x = p_x\bar{T}$, and the probability $\rho[T_x(\mathbb{I})] \equiv p_x$ for obtaining the measurement result x is independent of the input state ρ.*

2.6.3 Duality between Channels and Bipartite States

There are many connections between the properties of states on bipartite systems, and channels. For example, if Alice has locally created a state, and wants to send one half to Bob, the properties of the channel available for that transmission are crucial to the kind of distributed entangled state they can create in this way. For example, if the channel is separable, the state will also be separable.

Mathematically, the kind of relationship we shall describe here is very reminiscent of the relationship between bilinear forms and linear operators: an operator from an n-dimensional vector space to an m-dimensional vector space is parametrized by an $n \times m$ matrix, just like a bilinear form with arguments from an n-dimensional and an m-dimensional space. It is therefore hardly surprising that the matrix elements of a density operator on a tensor product can be reorganized and reinterpreted as the matrix elements of an operator between operator spaces. What is perhaps not so obvious, however, is that the positivity conditions for states and for channels exactly match up in this correspondence. This is the content of the following Lemma, graphically represented in Fig. 2.9.

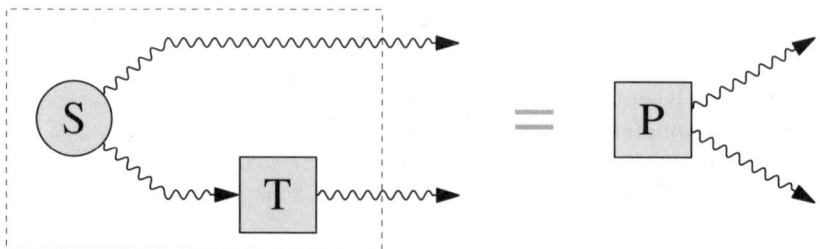

Fig. 2.9. The duality scheme of Lemma 2.2: an arbitrary preparation P is uniquely represented as a preparation S of a pure state and the application of a channel T to half of the system

Lemma 2.2. *Let ρ be a density operator on $\mathcal{H} \otimes \mathcal{K}$. Then there is a Hilbert space $\mathcal{H}'$, a pure state σ on $\mathcal{H} \otimes \mathcal{H}'$ and a channel $T : \mathcal{B}(\mathcal{K}) \to \mathcal{B}(\mathcal{H}')$ such that*

$$\rho = \sigma \circ (\mathbb{I}_\mathcal{H} \otimes T) \ . \tag{2.11}$$

Moreover, the restriction of σ to $\mathcal{H}'$ can be chosen to be nonsingular, and in this case the decomposition is unique in the sense that any other decomposition $\rho = \sigma' \circ (\mathbb{I}_\mathcal{H} \otimes T')$ is of the form $\sigma' = \sigma \circ R$ and $T' = R^{-1}T$, with a unitarily implemented channel R.

It is clear that σ must be the purification of ρ, restricted to the first factor. Thus we may set $\sigma = |\Psi\rangle\langle\Psi|$, where $\Psi = \sum_k \sqrt{r_k}\, e_k \otimes e'_k$; $r_k > 0$ are the nonzero eigenvalues of the restriction of ρ to the first system, and e'_k is a basis of $\mathcal{H}'$. Note that the e'_k are indeed unique up to a unitary transformation, so we only have to show that for one choice of e'_k we obtain a unique T. From the equation $\rho = \sigma \circ (\mathbb{I}_\mathcal{H} \otimes T)$, we can then read off the matrix elements of T:

$$\langle e'_k, T(|e_\mu\rangle\langle e_\nu|)\, e'_\ell\rangle = r_k^{-1/2} r_\ell^{-1/2} \rho\bigl(|e_k \otimes e_\mu\rangle\langle e_\ell \otimes e_\nu|\bigr) \ . \tag{2.12}$$

We have to show that T, as defined by this equation, is completely positive whenever ρ is positive. For fixed coefficients r_k the map $\rho \mapsto T$ is obviously linear. Hence it suffices to prove complete positivity for $\rho = |\varphi\rangle\langle\varphi|$. But in that case $T = V^*AV$, with $\langle e_\nu, Ve'_\ell\rangle = r_\ell^{-1/2}\langle e_\ell \otimes e_\nu, \varphi\rangle$, so T is indeed completely positive. The normalization $T(\mathbb{1}) = \mathbb{1}$ follows from the choice of r_k, and the lemma is proved.

The main use of this lemma is to translate results about entangled states to results about channels, and conversely. For this it is necessary to have a translation table of properties. Some entries are easy: for example, ρ is a product state iff T is *depolarizing* in the sense that $T(A) = \text{tr}(\rho_2 A)$ for some density operator ρ_2, and ρ is separable in the sense of Definition 2.1 iff T is separable (see (2.7)).

2.6.4 Channel Capacity

In the definition of channel capacity, we shall have to use a criterion for the approximation of one channel by another. Since channels are maps between normed spaces, one obvious choice would be to use the standard norm

$$\|S - T\| := \sup\{\|S(A) - T(A)\| \mid \|A\| \leq 1\}. \tag{2.13}$$

However, as in the case of positivity, there is a problem with this definition when one considers tensor products: the norms $\|T \otimes \mathbb{I}_n\|$, where $\mathbb{I}_n$ is the identity on $\mathcal{M}_n$, may increase with n. This introduces complications when one has to make estimates for parallel channels. Therefore we stabilize the norm with respect to tensoring with "innocent bystanders", and introduce, for any linear map T between C* algebras, the norm

$$\|T\|_{\text{cb}} := \sup_n \|T \otimes \mathbb{I}_n\|, \tag{2.14}$$

called the *norm of complete boundedness*, or "cb norm" for short. This name derives from the observation that on infinite-dimensional C* algebras the above supremum may be infinite even though each term in the supremum is finite. By definition, a completely bounded map is one with $\|T\|_{\text{cb}} < \infty$. On a finite-dimensional C* algebra, every linear map is completely bounded: for maps into $\mathcal{M}_d$ we have $\|T\|_{\text{cb}} \leq d\|T\|$. (As a general reference on these matters, I recommend the book [91].) One might conclude from this that the distinction between these norms is irrelevant. However, since we shall need estimates for large tensor products, every factor that increases with dimension can make a decisive difference. This is the reason for employing the cb norm in the definition of channel capacity. It will turn out, however, that in the most important cases one has only to estimate differences from the identity, and $\|T - \mathbb{I}\|$ and $\|T - \mathbb{I}\|_{\text{cb}}$ can be estimated in terms of each other with dimension-independent bounds.

The basis of the notion of channel capacity is a comparison between the given channel $T : \mathcal{A}_2 \to \mathcal{A}_1$ and an "ideal" channel $S : \mathcal{B}_1 \to \mathcal{B}_2$. The comparison is effected by suitable encoding and decoding transformations

$E: \mathcal{A}_1 \to \mathcal{B}_1$ and $D: \mathcal{B}_2 \to \mathcal{A}_2$ so that the composed operator $ETD: \mathcal{B}_2 \to \mathcal{B}_1$ is a map which can be compared directly with the ideal channel S. Of course, we are only interested in such a comparison in the case of optimal encoding and decoding, i.e. in the quantity

$$\Delta(S,T) = \inf_{E,D} \|S - ETD\|_{\text{cb}}, \qquad (2.15)$$

where the infimum is over all channels (i.e. all unit-preserving completely positive maps) E and D with appropriate domain and range. Since these data are at least implicitly given together with S and T, there is no need to specify them in the notation. S should be thought of as representing one word of the kind of message to be sent, whereas T represents one invocation of the channel. Channel capacity is defined as the number of S words per invocation of the channel T which can be faithfully transmitted, with suitable encoding and decoding for long messages. Here "messages of length n" are represented by the tensor power $S^{\otimes n}$, and "m invocations of the channel T" are represented by the tensor power $T^{\otimes m}$.

Definition 2.3. *Let S and T be channels. Then a number $c \geq 0$ is called an* achievable rate *for T with respect to S if, for any sequences n_α, m_α of integers with $m_\alpha \to \infty$ and $\limsup_\alpha (n_\alpha/m_\alpha) < c$, we have*

$$\lim_\alpha \Delta\bigl(S^{\otimes n_\alpha}, T^{\otimes m_\alpha}\bigr) = 0.$$

The supremum of all achievable rates is called the capacity *of T with respect to S, and is denoted by $C(S,T)$.*

Note that by definition, 0 is an achievable rate (no integer sequences with asymptotically negative ratio exist), and hence $C(S,T) \geq 0$. If all $c \geq 0$ are achievable, then of course we write $C(S,T) = \infty$. It may be cumbersome to check *all* pairs of integer sequences with a given upper ratio when testing c. However, owing to the monotonicity of Δ, it suffices to check only one sequence, provided it is not too sparse: if there is any pair of sequences n_α, m_α satisfying the conditions in the definition (including $\Delta \to 0$) plus the extra requirement that $(m_\alpha/m_{\alpha+1}) \to 1$, then c is achievable.

The *ideal channel* for systems with an algebra of observables $\mathcal{A}$ is by definition the identity map $\mathbb{I}_\mathcal{A}$ on $\mathcal{A}$. For typographical convenience we shall abbreviate "$\mathbb{I}_\mathcal{A}$" to "$\mathcal{A}$" whenever it appears as an argument of Δ or C. Using this notation, we shall now summarize the capacities of ideal quantum and classical channels. Of course, these are basic data for the whole theory:

$$C(\mathcal{M}_k, \mathcal{C}_n) = 0 \quad \text{for } k \geq 2, \qquad (2.16)$$

$$C(\mathcal{C}_k, \mathcal{C}_n) = C(\mathcal{M}_k, \mathcal{M}_n) = C(\mathcal{M}_k, \mathcal{C}_n) = \frac{\log n}{\log k}. \qquad (2.17)$$

Here the first equation is the capacity version of the no-teleportation theorem: it is impossible to transport any quantum information on a classical channel.

The second equation shows that for capacity purposes, $\mathcal{M}_n$ is indeed best compared with $\mathcal{C}_n$. In classical information theory one uses the one-bit system $\mathcal{C}_2$ as the ideal reference channel. Similarly, we use the one-qubit channel as the reference standard for quantum information, i.e. we define the *classical capacity* $C_c(T)$ and the *quantum capacity* $C_q(T)$ of an arbitrary channel by

$$C_c(T) = C(\mathcal{C}_2, T), \tag{2.18}$$
$$C_q(T) = C(\mathcal{M}_2, T). \tag{2.19}$$

Combining the results (2.17) with the "triangle inequality", or *two-step coding inequality*,

$$C(T_1, T_3) \geq C(T_1, T_2)C(T_2, T_3), \tag{2.20}$$

we see that this is really only a choice of units, i.e. for arbitrary channels T we obtain $C(\mathcal{M}_n, T) = (\log 2/\log n)C(\mathcal{M}_2, T)$, and a similar equation for classical capacities. Note that the term "qubit" refers to the reference system $\mathcal{M}_2$, but it is not advisable to use "qubit" as a special unit for quantum information (rather than just "bit"): this would be like distinguishing between the units "vertical meter" and "horizontal meter" and would create problems in every equation in which the two capacities were directly compared. The simplest relation of this kind is

$$C_q(T) \leq C_c(T), \tag{2.21}$$

which follows from combining (2.20) with (2.17). Note that both definitions apply to arbitrary channels T, whether the input and/or output are classical or quantum or hybrids. In order for a channel to have a positive quantum capacity, it is necessary that both the input and the output are quantum systems. This is shown by combining (2.16) with the *bottleneck inequality*

$$C(S, T_1 T_2) \leq \min\{C(S, T_1), C(S, T_2)\}. \tag{2.22}$$

Another application of the bottleneck inequality is to *separable channels*. These are by definition the channels with a purely classical intermediate stage, i.e. $T = SR$, where "output of S" = "input of R" is a classical system. For such channels $C_q(T) = 0$.

An important operation on channels is running two channels in parallel, represented mathematically by the tensor product. The relevant inequality is

$$C(S, T_1 \otimes T_2) \geq C(S, T_1) + C(S, T_2) \tag{2.23}$$

for the standard ideal channels, and when all systems involved are classical, we even have equality. However, it is one of the big unsolved problems to decide under what general circumstances this is true.

Comparison with the Classical Definition Since the definition of classical capacity $C_c(T)$ also applies to the purely *classical* situation, we have to verify that it is indeed equivalent to the standard definition in this case. To that end, we have to evaluate the error quantity $\|T - \mathbb{I}\|_{cb}$ for a classical-to-classical channel. As noted, a classical channel $T : \mathcal{C}(Y) \to \mathcal{C}(X)$ is given by a transition probability matrix $T(x \to y)$. Since the cb norm coincides with the ordinary norm in the classical case, we obtain

$$\|\mathbb{I} - T\|_{cb} = \|\mathbb{I} - T\| = \sup_{x,f} \left| \sum_y \bigl(\delta_{xy} - T(x \to y)\bigr) f(y) \right|$$
$$= 2 \sup_x \bigl[1 - T(x \to x)\bigr],$$

where the supremum is over all $f \in \mathcal{C}(Y)$ with $|f(y)| \leq 1$ and is attained where f is just the sign of the parenthesis in the second line, and we have used the normalization of the transition probabilities. Hence, apart from an irrelevant factor of two, $\|T - \mathbb{I}\|_{cb}$ is just the *maximal probability of error*, i.e. the largest probability for sending x and obtaining anything different. This is precisely the quantity which is required to go to zero (after suitable coding and decoding) in Shannon's classical definition of the channel capacity of discrete memoryless channels [92]. Hence the above definition agrees with the classical one.

When considering the classical capacity $C_c(T)$ of a quantum channel, it is natural to look at a coded channel ETD as a channel in its own right. Since we are considering transmission of classical information, this is a purely classical channel, and we can look at its classical capacity. Optimizing over coding and decoding, we obtain the quantity

$$C_{c,1}(T) = \sup_{ETD \text{ classical}} C_c(ETD). \tag{2.24}$$

This is called the *one-shot classical capacity*, because it can be said to involve only one invocation of the channel T. Of course, many uses of the channel are implicit in the capacity on the right-hand side, but these are in some sense harmless. In fact, every coding and decoding scheme for comparing $(ETD)^{\otimes n}$ with an ideal classical channel is also a coding/decoding for $T^{\otimes n}$, but the codings/decodings that arise in this way from the coding ETD are only those in which the coded input states and the measurements at the outputs are *not entangled*. If we allow entanglement over blocks of a large length ℓ, we thus recover the full classical capacity:

$$C_{c,1}(T) \leq C_c(T) = \sup_\ell \frac{1}{\ell} C_{c,1}(T^{\otimes \ell}). \tag{2.25}$$

It is not clear whether equality holds here. This is a fundamental question, which can be paraphrased as follows: "Does entangled coding ever help in sending classical information over quantum channels?" At the moment, all partial results known to the author seem to suggest that this is not the case.

Comparison with other Error Criteria Coming now to the quantum capacity $C_q(T)$, we have to relate our definition to more current definitions. One version, first stated by Bennett, is very similar to the one given above, but differs slightly in the error quantity, which is required to go to zero. Rather than $\|T - \mathbb{I}\|_{\mathrm{cb}}$, he considers the lowest *fidelity* of the channel, defined as

$$\mathcal{F}(T) = \inf_\psi \langle \psi, T(|\psi\rangle\langle\psi|)\psi\rangle , \qquad (2.26)$$

where the supremum is over all unit vectors. Hence the achievable rates are those for which $\mathcal{F}(ET^{\otimes n_\alpha} D) \to 1$, where E, D map to a system of m_α qubits, and these integer sequences satisfy the same constraints as above. This definition is equivalent to ours, because the error estimates are equivalent. In fact, if we introduce the *off-diagonal fidelity*

$$\mathcal{F}_\%(T) = \sup_{\phi,\psi} \Re\langle \phi, T(|\phi\rangle\langle\psi|)\psi\rangle \qquad (2.27)$$

for any channel $T : \mathcal{M}_d \to \mathcal{M}_d$ with $d < \infty$, we have the following system of estimates:

$$\|T - \mathbb{I}\| \leq \|T - \mathbb{I}\|_{\mathrm{cb}} \leq 4\sqrt{1 - \mathcal{F}_\%(T)} \leq 4\sqrt{\|T - \mathbb{I}\|}, \qquad (2.28)$$

$$\|T - \mathbb{I}\| \leq 4\sqrt{1 - \mathcal{F}(T)} \leq 4\sqrt{1 - \mathcal{F}_\%(T)} , \qquad (2.29)$$

which will be proved elsewhere. The main point is, though, that the dimension does not appear in these estimates, so if one such quantity goes to zero, all others do, and we can build an equivalent definition of capacity out of any one of them.

Yet another definition of quantum capacity has been given in terms of entropy quantities [93], and has also been shown to be equivalent [94].

2.6.5 Coding Theorems

The definition of channel capacity looks simple enough, but computing it on the basis of this definition is in general a very hard task: it involves an optimization over all coding and decoding channels in systems of asymptotically many tensor factors. Hence it is crucial to obtain simpler expressions which can be computed in a much more direct way from the matrix elements of the given channel. Such results are called *coding theorems*, after the first theorem of this type, established by Shannon.

In order to state this theorem, we need some entropy quantities. The *von Neumann entropy* of a state with density matrix ρ is defined as

$$S(\rho) = -\mathrm{tr}(\rho \log \rho) , \qquad (2.30)$$

where the function of ρ on the right-hand side is evaluated in the functional calculus, and $0 \log 0$ is defined to be zero. The logarithm is chosen here as the logarithm to base 2, so the unit of entropy is a "bit". The *relative entropy* of a state ρ with respect to another, σ, is defined by

$$S(\rho, \sigma) = \text{tr}\big[\rho(\log \rho - \log \sigma)\big]. \tag{2.31}$$

Both quantities are positive, and may be infinite on an infinite-dimensional space. The von Neumann entropy is concave, whereas the relative entropy is convex jointly in both arguments. For more precise definitions and many further results, I recommend the book by Ohya and Petz [95].

The strongest coding theorem for quantum channels known so far is the following expression for the one-shot classical capacity, proved by Holevo [96]:

$$C_{c,1}(T) = \max \left[S\Big(\sum_i p_i T_*[\rho_i]\Big) - \sum_i p_i S(T_*[\rho_i]) \right]. \tag{2.32}$$

Whether or not this is equal to the classical capacity depends on whether the conjectured equality in (2.25) holds or not. In any case, equality is known to hold for channels with classical input, so Holevo's coding theorem is a genuine extension of Shannon's.

No coding theorem has been proved yet for the quantum capacity. However, there is a fairly good candidate for the right-hand side, related to a quantity called "coherent information" [97]. The formula is written most compactly by relating it to an entanglement quantity via Lemma 2.2. For any bipartite state ρ with restriction ρ^B to the second factor, let

$$E_S(\rho) = S(\rho^B) - S(\rho). \tag{2.33}$$

This is a measure of entanglement of sorts, because it is large when $S(\rho)$ is small, e.g. when ρ is pure, and ρ^B is very mixed when, for example, ρ is maximally entangled. It can be negative, though (see [98] for a discussion). We set

$$C_{S,1}(T) = \sup_\sigma E_S[\sigma \circ (\mathbb{I} \otimes T)], \tag{2.34}$$

where the supremum is over all bipartite pure states σ. Note that any measure of entanglement can be turned into a capacity-like expression by this procedure. Since this quantity is known not to be additive [99], the candidate for the right-hand side of the quantum coding theorem is

$$C_S(T) = \sup_\ell \frac{1}{\ell} C_{S,1}(T^{\otimes \ell}), \tag{2.35}$$

in analogy to (2.25). So far there have been some good heuristic arguments [100, 101] in favor of this candidate, but a full proof remains one of the main challenges in the field.

An interesting upper bound on $C_q(T)$ can be written in terms of the transpose operation Θ on the output system [81]: we have

$$C_q(T) \leq \log_2 \|\Theta T\|_{\text{cb}} . \tag{2.36}$$

Hence, if ΘT happens to be completely positive (as for any channel with an intermediate classical state), this map is a channel; hence, it has a cb norm of 1, and $C_q(T) = 0$. This criterion can also be used to show that whenever there is sufficiently high noise in a channel, it will have a quantum capacity of zero.

2.6.6 Teleportation and Dense Coding Schemes

In this section we shall show that entanglement-assisted teleportation and dense coding, as described in Sects. 2.4.3 and 2.4.4, really work.

Rather than going through the now standard derivations in the basic examples involving qubits, we shall use the structure assembled so far to reverse the question, i.e. we try to find the *most general* setup in which teleportation and dense coding work without errors. This will give some additional insights, and possibly some welcome flexibility when it comes to realizing these processes for systems larger than a qubit. The task as stated in this form is somewhat beyond the scope of this chapter, mainly because there are so many ways to waste resources, which do not necessarily have a compact characterization. So, in order to obtain a readable result, we look only at the "tight case" [102], in which resources are used, in a sense, optimally. By this we mean that all Hilbert spaces involved have the same finite but arbitrary dimension d (so we can take them all equal to $\mathcal{H} = \mathbb{C}^d$), and the classical channel distinguishes exactly $|X| = d^2$ signals.

For both teleportation and dense coding, the beginning of each transmission is the distribution of the parts of an entangled state ω between the sender Alice and the receiver Bob. Only then is Alice given the message she is supposed to send, which is a quantum state in the case of teleportation and a classical value in the case of dense coding. She codes this in a suitable way, and Bob reconstructs the original message by evaluating Alice's signal jointly with his entangled subsystem.

For *dense coding*, assume that $x \in X$ is the message given to Alice. She encodes it by transforming her entangled system with a channel T_x and sending the resulting quantum system to Bob, who measures an observable F jointly on Alice's particle and his. The probability for obtaining y as a result is then $\text{tr}\big[\omega(T_x \otimes \mathbb{I})(F_y)\big]$, where the "$\otimes \mathbb{I}$" expresses the fact that no transformation is applied to Bob's particle, while Alice applies T_x to hers. If everything works correctly, this expression has to be equal to 1 for $x = y$, and 0 otherwise:

$$\text{tr}\big[\omega(T_x \otimes \mathbb{I})(F_y)\big] = \delta_{xy} . \tag{2.37}$$

Let us take a similar look at *teleportation*. Here three quantum systems are involved: the entangled pair in state ω, and the input system given to Alice, in state ρ. Thus the overall initial state is $\rho \otimes \omega$. Alice measures an observable F on the first two factors, obtaining a result x, which is sent to Bob. Bob applies a transformation T_x to his particle, and makes a final measurement of an observable A of his choice. Thus the probability of Alice measuring x and of Bob obtaining a result "yes" on A is $\mathrm{tr}(\rho \otimes \omega)[F_x \otimes T_x(A)]$. Note that the tensor symbols in this equation refer to different splittings of the system ($1 \otimes 23$ and $12 \otimes 3$, respectively). Teleportation is successful if the overall probability of obtaining A, computed by summing over all possibilities x, is the same as for an ideal channel, i.e.

$$\sum_{x \in X} \mathrm{tr}(\rho \otimes \omega)[F_x \otimes T_x(A)] = \mathrm{tr}(\rho A) \ . \tag{2.38}$$

Surprisingly, in the tight case one obtains exactly the same conditions on ω, T_x, F_x for teleportation and for dense coding, i.e. a dense-coding scheme can be turned into a teleportation scheme simply by letting Bob and Alice swap their equipment. However, this symmetry depends crucially on the tightness condition, because teleportation schemes with $|X| > d^2$ signals are trivial to achieve, but $|X| > d^2$ is impossible for dense coding. Conversely, dense coding through a $d' > d$-dimensional channel is trivial to achieve, while teleportation of states with $d' > d$ dimensions (with the same X) is impossible.

Let us now give a heuristic sketch of the arguments leading to the necessary and sufficient conditions for (2.37) and (2.38) to hold. For full proofs we refer to [102]. A crucial ingredient in the analysis of the teleportation equation is the "no measurement without perturbation" principle from Lemma 2.1: the left-hand side of (2.38) is indeed such a decomposition, so each term must be equal to $\lambda_x \mathrm{tr}(\rho A)$ for all ρ, A. But we can carry this even further: suppose we decompose ω, F_x or T_x into a sum of (completely) positive terms. Then each term in the resulting sum must also be proportional to $\mathrm{tr}(\rho A)$. Hence any components of ω, T_x or F_x satisfy a teleportation equation as well (up to normalization). Similarly, the vanishing of the dense-coding equation for $x \neq y$ carries over to every positive summand in ω, T_x or F_x. Hence it is plausible that we must first analyze the case where all ω, F_x, T_x are "pure", i.e. have no nontrivial decompositions as sums of (completely) positive terms:

$$\omega = |\Omega\rangle\langle\Omega|, \tag{2.39}$$
$$F_x = |\Phi_x\rangle\langle\Phi_x|, \tag{2.40}$$
$$T_x(A) = U_x^* A U_x \ . \tag{2.41}$$

The further analysis will show that in the pure case any two of these elements determine the third via the teleportation or the dense-coding equation, so that in fact all components of ω (and correspondingly T_x or F_x) have to be proportional. Hence each of these has to be pure in the first place. For the present discussion, let us just assume purity in the form (2.39)–(2.41) from now on. Note that normalization requires that each U_x is unitary.

The second normalization condition, $\sum_x |\Phi_x\rangle\langle\Phi_x| = \sum_x F_x = \mathbb{1}$, has an interesting consequence in conjunction with the tightness condition: the vectors Φ_x live in a d^2-dimensional space, and there are exactly d^2 of them. This implies that they are orthogonal: since each vector Φ_x satisfies $\|\Phi_x\| \leq 1$, and $d = \text{tr}(\mathbb{1}) = \sum_x \|\Phi_x\|^2$, we must have $\|\Phi_x\| = 1$ for all x. Hence, in the sum $1 = \sum_x \langle \Phi_y, F_x \Phi_y \rangle$ the term $y = x$ is equal to 1, and hence the others must be be zero.

Now consider the term with index x in the teleportation equation and set $\rho = |\phi'\rangle\langle\phi|$ and $A = |\psi\rangle\langle\psi'|$. Then the trace splits into two scalar products, in which the variables ϕ, ϕ', ψ, ψ' can be chosen independently, which leads to an equation of the form

$$\langle \phi \otimes \Omega, \Phi_x \otimes (U_x^* \psi) \rangle = \lambda_x \langle \phi, \psi \rangle , \qquad (2.42)$$

for all ϕ, ψ, and to coefficients which must satisfy $\sum_x |\lambda_x|^2 = 1$. Note how in this equation a scalar product between the vectors in the first and third tensor factors is generated. This type of equation, which is clearly the core of the teleportation process, may be solved in general:

Lemma 2.3. *Let $\mathcal{H}, \mathcal{K}$ be finite-dimensional Hilbert spaces, and let $\Omega_1 \in \mathcal{K} \otimes \mathcal{H}$ and $\Omega_2 \in \mathcal{H} \otimes \mathcal{K}$ be unit vectors such that, for all $\phi, \psi \in \mathcal{H}$,*

$$\langle \phi \otimes \Omega_1, \Omega_2 \otimes \psi \rangle = \lambda \langle \phi, \psi \rangle . \qquad (2.43)$$

Then $|\lambda| \leq 1/\dim \mathcal{H}$, with equality iff Ω_1 and Ω_2 are maximally entangled and equal up to the exchange of the tensor factors $\mathcal{H}$ and $\mathcal{K}$.

For the proof, consider the Schmidt decomposition $\Omega_1 = \sum_k \sqrt{w_k} f_k \otimes e_k$, and insert $\phi = e_n$, $\psi = e_m$ into (2.43) to find the matrix elements of Ω_2:

$$\langle e_n \otimes f_m, \Omega_2 \rangle = \lambda \, w_m^{-1/2} \, \delta_{nm} .$$

Clearly, $\|\Omega_2\|^2 = |\lambda|^2 \sum_m w_m^{-1}$. This sum takes its smallest value under the constraint $\sum_m w_m = \|\Omega_1\|^2 = 1$ only at the point where all w_m are equal. This proves the lemma.

We apply this lemma to $\Omega_1 = (\mathbb{1} \otimes U_x)\Omega$ and $\Omega = \Phi_x$. Then $\sum_x |\lambda_x|^2 \leq d^2 d^{-2} = 1$, with equality only if all the vectors involved are maximally entangled and are pairwise equal up to an exchange of factors:

$$\Phi_x = (U_x \otimes \mathbb{1})\Omega , \qquad (2.44)$$

where we take $\Omega = d^{-1/2} \sum_k e_k \otimes e_k$ by an appropriate choice of bases. If Ω is maximally entangled, (2.44) sets up a one-to-one correspondence between unitary operators U_x and the vectors Φ_x, as independent elements in the construction. The Φ_x have to be an orthonormal basis of maximally entangled vectors, and there are no further constraints. In terms of the U_x, the orthogonality of the Φ_x translates into orthogonality with respect to the Hilbert–Schmidt scalar product:

$$\operatorname{tr}(U_x^* U_y) = d\,\delta_{xy} \ . \tag{2.45}$$

Again, there are no further constraints, so any collection of d^2 unitaries satisfying these equations leads to a teleportation scheme.

For the dense-coding case we obtain the same result, although by a different route. Equation (2.44) follows easily if we write the teleportation equation as $|\langle \Omega, (U_x^* \otimes \mathbb{1})\Phi_x \rangle|^2 = \delta_{xy}$. The problem is to show that Ω has to be maximally entangled. Using the reduced density operator ω_1 of ω, this becomes

$$\operatorname{tr}(\omega_1 U_x^* U_y) = \langle \Omega, (U_x^* U_y \otimes \mathbb{1})\Omega\rangle = \langle \Phi_x, \Phi_y\rangle = \delta_{xy} \ . \tag{2.46}$$

We claim that this equation, for a positive operator ω_1 and d^2 unitaries U_x, implies that $\omega_1 = d^{-1}\mathbb{1}$. To see this, expand the operator $A = |\phi\rangle\langle e_k|\omega_1^{-1}$ in the basis U_x according to the formula $A = \sum_x U_x \operatorname{tr}(U_x^* A \omega_1)$:

$$\sum_x \langle e_k, U_x^* \phi\rangle\, U_x = |\phi\rangle\langle e_k|\omega_1^{-1} \ .$$

Taking the matrix element $\langle\phi|\cdot|e_k\rangle$ of this equation and summing over k, we find

$$\sum_{x,k} \langle e_k, U_x^* \phi\rangle\,\langle\phi, U_x e_k\rangle = \sum_x \operatorname{tr}(U_x^*|\phi\rangle\langle\phi|U_x) = d^2 \|\phi\|^2 = \|\phi\|^2 \operatorname{tr}(\omega_1^{-1}) \ .$$

Hence $\operatorname{tr}(\omega_1^{-1}) = d^2 = \sum_k r_k^{-1}$, where r_k are the eigenvalues of ω_1. Using again the fact that the smallest value of this sum under the constraint $\sum_k r_k = 1$ is attained only for constant r_k, we find $\omega_1 = d^{-1}\mathbb{1}$, and Ω is indeed maximally entangled.

To summarize, we have the following theorem (again, for a detailed proof see [102]):

Theorem 2.3. *Given either a teleportation scheme or a dense-coding scheme, which is tight in the sense that all Hilbert spaces are d-dimensional and $|X| = d^2$ classical signals are distinguished, then*

- $\omega = |\Omega\rangle\langle\Omega|$ *is pure and maximally entangled,*
- $F_x = |\Phi_x\rangle\langle\Phi_x|$, *where the Φ_x form an orthonormal basis of maximally entangled vectors,*
- $T_x(A) = U_x^* A U_x$, *where the U_x are unitary and orthonormal in the sense that $\operatorname{tr}(U_x^* U_y) = d\,\delta_{xy}$, and*
- *these objects are connected by the equation $\Phi_x = (U_x \otimes \mathbb{1})\Omega$.*

Given either the Φ_x or the U_x with the appropriate orthogonality properties, and a maximally entangled vector Ω, the above conditions determine a dense coding and a teleportation scheme.

In particular, we have shown that a teleportation scheme becomes a dense-coding scheme and vice versa, when Alice and Bob swap their equipment. However, this is only true in the tight case: for a larger quantum channel, dense coding becomes easier but teleportation becomes more demanding. Similarly, teleportation becomes easier with more allowed classical information exchange, whereas dense coding of more than d^2 signals is impossible.

In order to construct a scheme, it is best to start from the equation $\text{tr}(U_x^* U_y) = d\, \delta_{xy}$, i.e. to look for orthonormal bases in the space of operators consisting of unitaries. For $d = 2$ the solution is essentially unique: $U_1, \ldots, U_4$ are the identity and the three Pauli matrices, which leads to the standard examples. Group theory helps to construct examples of such bases for any dimension d, but this construction by no means exhausts the possibilities. A fairly general construction is given in [102]. It requires two combinatorial structures known from classical design theory [103, 104]: a *Latin square* of order d, i.e. a matrix in which each row and column is a permutation of $(1, ..., d)$, and d *Hadamard matrices*, i.e. unitary $d \times d$ matrices, in which each entry has modulus $d^{-1/2}$. For neither Latin squares nor Hadamard matrices does an exhaustive construction exist, so these are rich fields for hunting and gathering new examples, or even infinite families of examples. Certainly, this connection suggests that a full classification or exhaustive construction of teleportation and dense-coding schemes cannot be expected. However, it may still be a good project to look for schemes with additional desirable features.

3 Quantum Communication

Harald Weinfurter and Anton Zeilinger

Quantum entanglement lies at the heart of the new field of quantum communication and computation. For a long time, entanglement was seen just as one of those fancy features which make quantum mechanics so counterintuitive. But recently, quantum information theory has shown the tremendous importance of quantum correlations for the formulation of new methods of information transfer and for algorithms exploiting the capabilities of quantum computers. While the latter needs entanglement between a large number of quantum systems, the basic quantum communication schemes rely only on entanglement between the members of a pair of particles, directly pointing to a possible realization of such schemes by means of correlated photon pairs such as those produced by parametric down-conversion.

This chapter describes the first experimental realizations of quantum communication schemes using entangled photon pairs. We show how to make communication secure against eavesdropping using entanglement-based quantum cryptography, how to increase the information capacity of a quantum channel by quantum dense coding and, finally, how to communicate quantum information itself in the process of quantum teleportation.

3.1 Introduction

Quantum mechanics is probably the most successful physical theory of this century. It provides powerful tools which form one of the cornerstones of scientific progress, and which are indispensable for the understanding of omnipresent technical devices such as the transistor, semiconductor chips and the laser. The most important areas where those devices are used are modern communication and information-processing technologies. But quantum mechanics, until now, has only been used to construct these devices – quantum effects are absolutely avoided in the representation and manipulation of information. Rather than using single photons, one still uses strong light pulses to send information along optical high-speed connections, and one relies on electrical currents in semiconductor logic chips instead of applying single electrons as signal carriers.

This caution surely is due to the fact that, at first glance, the inherent stochastic character of quantum effects seems only to introduce unavoidable noise and thus does not really recommend their use. Yet quantum information theory shows us, in more and more examples, how one can profit from the peculiar properties of quantum systems, and, when applied correctly, how fundamental quantum effects can add to the power and features of classical information processing and transmission [12, 105, 106]. For example, quantum computers will outperform conventional computers, and quantum cryptography enables, for the first time, secure communication. While quantum cryptography, in principle, can be performed even with single quantum particles, all the other proposals utilize entanglement between two or more particles, for example to enhance communication rates or to enable the teleportation of quantum states.

Entanglement between quantum systems is a pure quantum effect. It is closely related to the superposition principle and describes correlations between quantum systems that are much stronger and richer than any classical correlation could be. Originally this property was introduced by Einstein, Podolsky and Rosen (EPR) [24], and also by Schrödinger [5] and Bohr [107] in the discussion of the completeness of quantum mechanics and by von Neumann [108] in his description of the measurement process. Entanglement also provides a handle to distinguish various interpretations of quantum mechanics via Bell's theorem [72, 109] or the GHZ argument [37]. The development of experimental techniques has enabled researchers to perform the recent long-distance tests of entanglement [30], the first Bell experiment fulfilling Einstein locality conditions [31] and the first GHZ experiment [38], which all provided convincing demonstrations of the validity of standard quantum mechanics.[1]

The field of quantum information is not concerned with the fundamental issues. Instead, it builds on the validity of quantum mechanics and applies the characteristic features of entangled systems to devise new, powerful schemes for communication and computation. Entanglement between a large number of quantum systems will enable very efficient computations. In particular, the factorization algorithm of Shor [9] and the search algorithm of Grover [58] (together with the increasing number of algorithms derived from one or the other) show how entanglement and the associated interference between entangled states can boost the power of quantum computers.

Quantum communication exploits entanglement between only two or three particles. As will be seen in the following sections, the often counterintuitive features of such small entangled systems enable powerful communication methods. After the very basic properties of pairs of entangled particles have been introduced (Sect. 3.2), Sect. 3.3 gives an overview of the possibilities of three important quantum communication schemes: entanglement-based quantum cryptography enables secret key exchange and thus truly secure communication [46]; using quantum dense coding, one can send classical in-

---

[1] We are aware of the detection loophole [110], which will be closed whenever technology allows.

formation more efficiently [51]; and, finally, with quantum teleportation one can transfer quantum information, that is, the quantum state itself, from one quantum system to another [52]. The tools for the experimental realization of those quantum communication schemes are presented in Sect. 3.4. In particular, we show how to produce polarization-entangled photon pairs by parametric down-conversion [111] and how to observe these nonclassical states by interferometric Bell-state analysis [112]. In Sect. 3.5 we describe the first experimental realizations of basic quantum communication schemes. In experiments performed during recent years at the University of Innsbruck, we could realize entanglement-based quantum cryptography with randomly switched analyzers and with the two users separated by more than 400 m [113]; we demonstrated the possibility of transmitting 1.58 bits of classical information by encoding trits on a single two-state photon [114]; and we could transfer a qubit, in our case the polarization state, from one photon to another by quantum teleportation [10, 11] and entanglement swapping [115].

3.2 Entanglement – Basic Features

For a long time, entanglement was seen merely as one of the counterintuitive features of quantum mechanics, important only within the realm of the EPR paradox. Only lately has the field of quantum information exploited these features to obtain new types of information transmission and processing. Recent literature now offers a thorough discussion of all the various properties of entangled systems [37, 72, 116] (see also Chap. 5); in this review, we concentrate on those features which form the foundation of the basic quantum communication schemes.

At the heart of entanglement lies another fundamental feature of quantum mechanics, the superposition principle. If we look at a classical, two-valued system, for example a coin, we find it in either one of its two possible states, that is, either head or tail. Its quantum mechanical counterpart, a two-state quantum system, however, can be found in any superposition of two possible basis states, e.g. $|\Psi\rangle = (1/\sqrt{2})(|0\rangle + |1\rangle)$. Here we denote the two orthogonal basis states by $|0\rangle$ and $|1\rangle$, respectively.[2] This generic notation can stand for any of the properties of various two-state systems, for example for the ground state $|g\rangle$ and excited state $|e\rangle$ of an atom, or, as is the case in our experiments, for the horizontal polarization $|H\rangle$ and vertical polarization $|V\rangle$ of a photon.

In the classical world, we find two coins to be in the states of either head/head, head/tail, tail/head or tail/tail, and we can identify these four possibilities with the four quantum states

$$|0\rangle_1|0\rangle_2,\ |0\rangle_1|1\rangle_2,\ |1\rangle_1|0\rangle_2\ \text{and}\ |1\rangle_1|1\rangle_2\,,$$

---

[2] This notation should not be confused with the description of an electromagnetic field (vacuum or single-photon state) in second quantization. Here we use only the notions of first quantization to describe the properties of two-state systems.

describing two two-state quantum systems. As the superposition principle holds for more than one quantum system, the two quantum particles are no longer restricted to the four "classical" basis states above, but can be in any superposition thereof, for example in the entangled state

$$|\Psi\rangle = \frac{1}{\sqrt{2}}(|0\rangle_1|0\rangle_2 + |1\rangle_1|1\rangle_2) \,. \tag{3.1}$$

Of course, one is restricted neither to only two particles nor to such maximally entangled states. During the last decade, enormous progress was achieved in the theoretical studies of the quantum features of multiparticle systems. One will observe even more stunning correlations between three or more entangled particles [37,117]; one can generalize to the observation of interference and entanglement between multistate particles [118] and to entanglement for mixed states. There also exists the possibility to purify entanglement [119], and one can even find two-particle systems which are not actually entangled, but are such that a local observer cannot distinguish them from entangled states [120]. For the basic quantum communication schemes and experiments, we can concentrate on the particular properties of maximally entangled two-particle systems. Considering two two-state particles, we find a basis of four orthogonal, maximally entangled states, the so called Bell-states basis:

$$|\Psi^+\rangle_{12} = \frac{1}{\sqrt{2}}(|0\rangle_1|1\rangle_2 + |1\rangle_1|0\rangle_2), \tag{3.2}$$

$$|\Psi^-\rangle_{12} = \frac{1}{\sqrt{2}}(|0\rangle_1|1\rangle_2 - |1\rangle_1|0\rangle_2), \tag{3.3}$$

$$|\Phi^+\rangle_{12} = \frac{1}{\sqrt{2}}(|0\rangle_1|0\rangle_2 + |1\rangle_1|1\rangle_2), \tag{3.4}$$

$$|\Phi^-\rangle_{12} = \frac{1}{\sqrt{2}}(|0\rangle_1|0\rangle_2 - |1\rangle_1|1\rangle_2) \,. \tag{3.5}$$

The name "Bell states" was given to these states since they maximally violate a Bell inequality [121]. This inequality was deduced in the context of so-called local realistic theories (see Chap. 1), and gives a range of possible results for certain statistical tests on identically prepared pairs of particles [109]. Quantum mechanics predicts different results if the measurements are performed on entangled pairs. If the two particles are not correlated, i.e. are described by a product state, the quantum mechanical prediction is within the range given by Bell's inequality.

The remarkably nonclassical features of entangled pairs arise from the fact that the two systems can no longer be seen as being independent but now have to be seen as one combined system, where the observation of one of the two will change the possible predictions of measurement results obtained for the other [5,107]. Formally, this mutual dependence is reflected by the fact that the entangled state can no longer be factored into a product of two states for the two subsystems separately.

If one looks only at one of the two particles, one finds it with equal probability in state $|0\rangle$ or in state $|1\rangle$. One has no information about the particular outcome of a measurement to be performed. However, the observation of one of the two particles determines the result of a measurement of the other particle. This holds not only for a measurement in the basis $|0\rangle/|1\rangle$, but for any arbitrary superposition, that is, for any arbitrary orientation of the measurement apparatus. In particular, for the state $|\Psi^-\rangle$ we shall find the two particles always in orthogonal states, no matter which measurement apparatus is used. If, for the case of polarization-entangled photons, we observe only one of the two photons, it appears to be completely unpolarized, and any polarization direction is observed with equal probability. However, the results for both photons are perfectly correlated. For example, this means that photon 2 has vertical polarization if we found horizontal polarization for photon 1, but also that photon 2 will be circularly polarized left if we observed right circular polarization for photon 1.

Another important feature of the four Bell states is that a manipulation of only one of the two particles suffices to transform from any Bell state to any of the other three states. This is not possible for the basis formed by the products. For example, to transform $|0\rangle_1|0\rangle_2$ into $|1\rangle_1|1\rangle_2$ one has to flip the state of both particles.

These three features,

- different statistical results for measurements on entangled or unentangled pairs
- perfect correlations between the observations of the two particles of a pair, although the results of the measurements on the individual particles are fully random
- the possibility to transform between the Bell states by manipulating only one of the two particles,

are the ingredients of the fundamental quantum communication schemes described here.

3.3 The Quantum Communication Schemes

Quantum communication methods utilize fundamental properties of quantum mechanics to enhance the power and potential of today's communication systems. The first step towards quantum information processing is the generalization of classical digital encoding, which uses the bit values "0" and "1". Quantum information associates two distinguishable, orthogonal states of a two-state system with these bit values. We thus directly translate the two values of a classical bit to the two basis states $|0\rangle$ and $|1\rangle$.

As an extension to the situation for classical communication, the quantum system can be in any superposition of the two basis states. To distinguish such a quantum state and the information contained in it from a classical bit, such a state is called a "qubit" [69]. The general state of a qubit is

$$|\Psi\rangle = a_0|0\rangle + a_1|1\rangle, \tag{3.6}$$

where a_0 and a_1 are complex amplitudes (with $|a_0|^2 + |a_1|^2 = 1$).

A measurement of the qubit projects the state onto either $|0\rangle$ or $|1\rangle$ and therefore cannot give the full quantum information about the state. Evidently, if we want to communicate information, we have to restrict ourselves to sending only basis states in order to avoid errors, and thus only one bit of classical information can be sent with a single qubit. Therefore, the new features do not seem to offer additional power. However, by provoking errors, quantum cryptography [42, 122] enables one to check the security of quantum key generation. The security of quantum cryptography relies on the fact that an eavesdropper cannot unambiguously read the state of a single quantum particle which is transferred from Alice to Bob.

When two-particle systems are used, entanglement adds many more features to the capabilities of quantum communication systems compared with classical systems. In recent years, several proposals have shown how to exploit the basic features of entangled states in new quantum communication schemes. In the following we shall see how entangled pairs enable a new formulation of quantum cryptography (Sect. 3.3.1), how we can surpass the limit of transmitting only one bit per qubit (Sect. 3.3.2) and how entanglement allows one to transfer quantum information from one particle to another in the process of quantum teleportation (Sect. 3.3.3).

3.3.1 Quantum Cryptography

Suppose two parties, let us call them Alice and Bob, want to send each other secret messages. There exists a cryptographic method, the one-time pad scheme,[3] which is secure against eavesdropping attacks – provided the key used for encoding and decoding the message is perfectly random, is as long as the original message and, most importantly, is secret and known only to Alice and Bob. But how can they be sure that the key was securely distributed to the two, and that no third person has knowledge about the key? Quantum cryptography [42, 122] provides a means to ensure the security of

---

[3] In the so-called "one-time pad" encryption (see Sect. 1), every character of the message is encoded with a random key character. As shown by Shannon [44], the cipher cannot be decoded without a knowledge of the key. The eavesdropping is impossible as long as the key is securely exchanged between the sender and receiver.

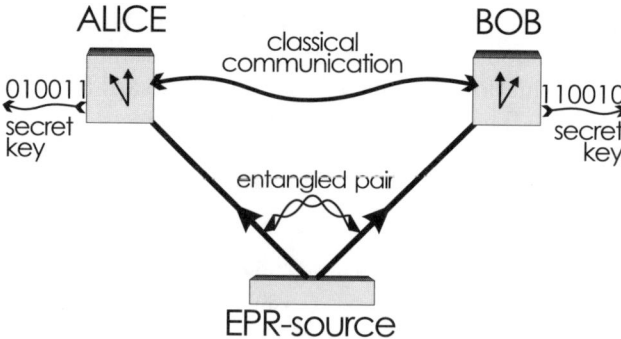

Fig. 3.1. Scheme for entanglement-based quantum cryptography [46]

the key distribution and thus enables, together with the one-time pad scheme, absolutely secret communication.[4]

Let us first discuss how quantum cryptography can profit from the fascinating properties of entangled systems to provide secure key exchange [46, 123]. Suppose that Alice and Bob receive particles which are in entangled pairs, from an EPR source (Fig. 3.1). Beforehand, Alice and Bob agreed on some preferred basis, again called $|0\rangle/|1\rangle$, in which they start to perform measurements. The possible results, $+1$ and -1, correspond to observation of the state $|1\rangle$ or $|0\rangle$, respectively. Owing to the entanglement of the particles, the measurement results of Alice and Bob will be perfectly correlated or, in a case where the source produces pairs in the $|\Psi^-\rangle$ state, perfectly anticorrelated. For each instance where Alice obtained -1, she knows that Bob observed $+1$, and if she obtained the result $+1$, she knows that Bob had -1. Alice and Bob can translate the result -1 to the bit value 0 and the result $+1$ to the bit value 1 and thereby establish a random key, ideal for encoding messages. But how can they be sure that no eavesdropper has intercepted the key exchange? There are two different techniques. The first scheme for entanglement-based quantum cryptography [123] builds on the ideas of the basic quantum cryptography protocol for single photons [42, 122]. In this case, Alice and Bob randomly and independently vary their analysis directions between $0°$, corresponding to the $|0\rangle/|1\rangle$ basis, and $45°$, corresponding to a second, noncommuting basis. They will observe perfect anticorrelations of their measurements whenever they happen to have polarizers oriented parallel (Alice and Bob thus obtain identical keys, if one of them inverts all bits of his/her key string). This can be viewed in the following way: as Alice makes

---

[4] For descriptions of quantum cryptography schemes not relying on entanglement, see [105, 106].

a measurement on photon A she projects photon B into the orthogonal state, which is then analyzed by Bob. An eavesdropper, not knowing the actual basis, causes errors, since he/she cannot determine the quantum state without information about the basis. Thus, Alice and Bob can find out, by communication over a classical, public channel, whether or not their key exchange has been attacked by checking whether or not some of the key bits are different. Of course, those test key bits cannot be used for secure communication and have to be sacrificed.

The other technique uses the fragility of entanglement against measurements. Any attack an eavesdropper might perform reduces the entanglement and allows Alice and Bob to check the security of their quantum key exchange. As described in Sect. 3.2, measurements on entangled pairs obey statistical correlations and will violate a Bell inequality. It can be shown that the more knowledge the eavesdropper has gained when he/she intercepted the key exchange, the less the inequality is violated. The amount by which a Bell inequality is violated is thus an ideal measure of the security of the key. Alice and Bob therefore measure the entangled particles not only in the basis $|0\rangle/|1\rangle$, but also along some other directions, depending on the Bell inequality used. A particularly simple form of Bell inequality, which is well suited for experimental application, is the version deduced by Wigner [28], which can be used as follows.

Alice chooses between two polarization measurements of photon A, either along the axis α or along the axis β, and Bob chooses between measurements along β and γ of photon B. We identify the direction β, which is common to the two users, with our preferred basis $|0\rangle/|1\rangle$. A detected polarization parallel to the analyzer axis corresponds to a $+1$ result, and a polarization orthogonal to the analyzer axis corresponds to -1. If, heretically, one assumes that every photon carries preassigned values determining the outcomes of the measurements on each of the photon pairs, it follows that the probabilities of obtaining $+1$ on both sides, p_{++}, must obey Wigner's inequality:

$$p_{++}(\alpha_A, \beta_B) + p_{++}(\beta_A, \gamma_B) - p_{++}(\alpha_A, \gamma_B) \geq 0 \,. \tag{3.7}$$

The quantum mechanical prediction p_{++}^{qm} for these probabilities with some arbitrary analyzer settings Θ_A (Alice) and Θ_B (Bob) and measurement of the Ψ^- state is

$$p_{++}^{qm}(\Theta_A, \Theta_B) = \frac{1}{2} \sin^2(\Theta_A - \Theta_B) \,. \tag{3.8}$$

The analyzer settings $\alpha = -30°$, $\beta = 0°$ and $\gamma = 30°$ lead to a maximum violation of Wigner's inequality (3.7):

$$\begin{aligned} & p_{++}^{qm}(-30°, 0°) + p_{++}^{qm}(0°, 30°) - p_{++}^{qm}(-30°, 30°) \\ & = \frac{1}{8} + \frac{1}{8} - \frac{3}{8} = -\frac{1}{8} \,, \end{aligned} \tag{3.9}$$

which is not greater than or equal to 0.

In order to implement quantum key distribution, Alice and Bob each vary their analyzers randomly between two settings: Alice uses $-30°, 0°$, and Bob uses $0°, 30°$. Because Alice and Bob operate independently, four possible combinations of analyzer settings will occur, of which the three oblique settings allow a test of Wigner's inequality and the remaining combination of parallel settings allows the generation of keys via the perfect anticorrelation (where, again, either Alice or Bob has to invert all bits of the key to obtain identical keys). If the measured probabilities violate Wigner's inequality, the security of the quantum channel is ascertained, and the keys generated can readily be used. This scheme is an improvement on the Ekert scheme, which uses the CHSH inequality. Since there are fewer settings on each side, the above version is technically easier to implement and also uses the photon pairs more efficiently for key generation.

Compared with standard attenuated-pulse quantum cryptography, such systems are practically immune to any beam-splitter attack (or other attacks that try to split pulses containing more than one photon) by a potential eavesdropper. First of all, a photon pair source can be used as an (almost) ideal source of single photons. If one of the photons is detected, the gate time of the coincidence electronics (typically on the order of 1 ns) determines the equivalent pulse duration in standard quantum cryptography. Since the probability of generating one photon pair during such a short time is very low, e.g., for the experiment described in Sect. 3.5.1, only about 6.8×10^{-4}, the probability of having two photons in the gate time is less than 3×10^{-7} and can be almost neglected. This has to be compared with a probability of having two photons in a pulse of 0.005 for a typical quantum cryptography realization using a mean of 0.1 photons per pulse.

However, the security against beam-splitting attacks can be further increased when entanglement-based schemes are used. In this case, there is only a correlation between two entangled pairs if they are simultaneously generated during a time interval of the order of the coherence time of the photons, i.e. during a time of typically 500 fs. This reduces the chances of an eavesdropper learning the value of a key bit to about 6×10^{-14} and guarantees unprecedented security of the quantum key. Moreover, by utilizing the peculiar properties of entangled photon pairs produced by parametric down-conversion, one immediately profits from the inherent randomness of quantum mechanical observations, which guarantees a truly random and non-deterministic key.

3.3.2 Quantum Dense Coding

If one wants to send some information, one encodes the message with distinguishable symbols, writes them on some physical entity and finally, this is transmitted to the receiver. To send one bit of information one uses, for example, the binary values "0" and "1" as code symbols written on the information

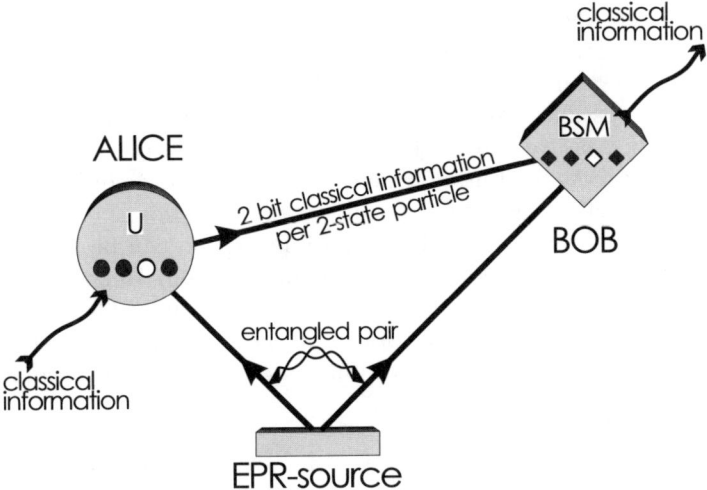

Fig. 3.2. Scheme for the efficient transmission of classical information by quantum dense coding [51] (BSM, Bell-state measurement; U, unitary transformation)

carrier. If one wants to send two bits of information, one consequently has to perform the process twice; that means one has to send two such entities.

As mentioned above, in the case of quantum information one identifies the two binary values with the two orthogonal basis states $|0\rangle$ and $|1\rangle$ of the qubit. In order to send a classical message to Bob, Alice uses quantum particles, all prepared in the same state by some source. Alice translates the bit values of the message by either leaving the state of the qubit unchanged or flipping it to the other, orthogonal state, and Bob, consequently, will observe the particle in one or the other state. That means that Alice can encode one bit of information in a single qubit. Obviously, she cannot do better, since in order to avoid errors, the states arriving at Bob have to be distinguishable, which is only guaranteed when orthogonal states are used. In this respect, they do not gain anything by using qubits as compared with classical bits. Also, if she wants to communicate two bits of information, Alice has to send two qubits.

Bennett and Wiesner found a clever way to circumvent the classical limit and showed how to increase the channel capacity by utilizing entangled particles [51]. Suppose the particle which Alice obtained from the source is entangled with another particle, which was sent directly to Bob (Fig. 3.2). The two particles are in one of the four Bell states, say $|\Psi^-\rangle$. Alice now can use a particular feature of the Bell basis, that manipulation of one of the two entangled particles suffices to transform to any other of the four Bell states. Thus she can perform one out of *four* possible transformations – that is, doing nothing, shifting the phase by π, flipping the state, or flipping and phase-shifting the state – to transform the two-particle state of their common pair to another state. After Alice has sent the transformed two-state particle to Bob, he can read the information by performing a combined measurement on both particles. He makes a measurement in the Bell-state basis and can identify which of the four possible messages was sent by Alice. Thus it is possible to encode two bits of classical information by manipulating and transmitting a single two-state system. Entanglement enables one to communicate information more efficiently than any classical system could do.

The preceding examples show how quantum information can be applied for secure and efficient transmission of classical information. But can one also transmit quantum information, that is, the state of a qubit? Obviously, quantum mechanics places a number of obstacles in the way of this intention, above all, the problem of measuring quantum states, which is utilized in quantum cryptography as already described.

3.3.3 Quantum Teleportation

The Idea It is an everyday task, in our classical world, for Alice to send some information to Bob. Imagine a fax machine. Alice might have some message, written on a sheet of paper. For the fax machine the actual written information does not matter, in fact, it reduces to just a sequence of white and black pixels. For the transmission, the machine scans the paper pixel by pixel. It measures whether a pixel is white or black and sends this information to Bob's machine, which writes the state of each pixel onto another sheet of paper. In classical physics, by definition, one can make the measurements with arbitrary precision, and Bob's sheet can thus become an ideal copy of Alice's original sheet of paper. If Alice's pixels were made smaller and smaller, they would, in reality, sooner or later be encoded on single molecules or atoms. If we again confined ourselves to coding in only the basis states, we surely could measure and transfer the binary value of even such pixels.

Now, imagine Alice not only has classical binary values encoded on her system, but wants to send a quantum state, i.e. quantum information, to Bob. She has a qubit encoded on some quantum system such as a molecule or atom, and wishes, that a quantum system in Bob's hands should represent this qubit at the end of the transmission. Evidently, Alice cannot read the quantum information, that is, measure the state of the quantum object with arbitrary precision. All she would learn from her measurement would be that the amplitude of the observed basis state was not zero. But this is not enough information for Bob to reconstruct the qubit on his quantum particle.

Another limitation, which definitely seems to bring the quest for perfect transfer of the quantum information to an end, is the no-cloning theorem (see Sect. 1) [4]. According to this theorem, the state of a quantum system cannot be copied onto another quantum system with arbitrary precision. Thus, how could Bob's quantum particle obtain the state of Alice's particle?

In 1993 Bennett et al. found the solution to this problem [52]. In their scheme, a chain of quantum correlations is established between the particle

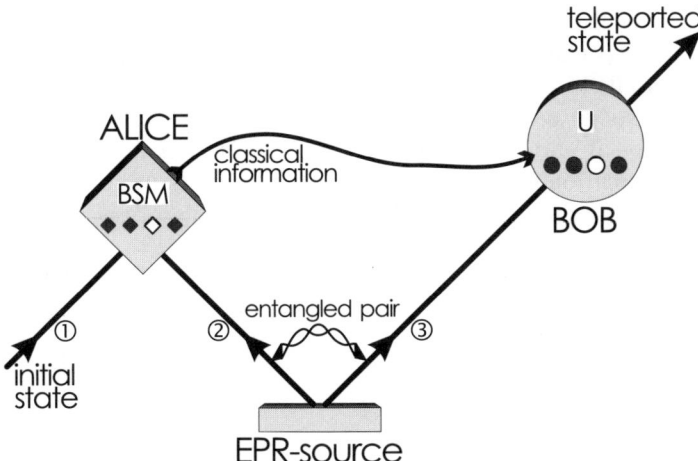

Fig. 3.3. Scheme for teleporting a quantum state from one system to another one [52]

carrying the initial quantum state and Bob's particle. They dispense with measuring the initial state; actually, they avoid gaining any knowledge about this state at all!

To perform quantum teleportation, initially Alice and Bob share an entangled pair of particles 2 and 3, which they have obtained from some source of entangled particles, in, say, the state $|\Psi^-\rangle_{2,3}$ (Fig. 3.3). As mentioned before, we cannot say anything about the state of particle 2 on its own. Nor do we know the state of particle 3. In fact, these particles do not have a (pure) state at all. But, whatever the results of measurements might be, we know for sure that they are orthogonal to each other. Next, particle 1, which carries the state to be sent to Bob, is given to Alice. She now measures particle 1 and 2 together, by projecting them onto the Bell-state basis. After projecting the two particles into an entangled state, she cannot infer anything about the individual states of particles 1 and 2 anymore. However, she knows about correlations between the two. Let us assume she has obtained the result $|\Psi^-\rangle_{1,2}$. This tells her, that whatever the two states of particles 1 and 2 have been, they have been orthogonal to each other. But from this, Alice already knows that the state of particle 3 is equal to the state of particle 1 (up to a possible overall phase shift). This follows because the state of particle 1 was orthogonal to 2 and, owing to the preparation of particles 2 and 3, the state of particle 2 was orthogonal to 3. All Alice has to do is to tell this to Bob, to let him know that, in this particular case, the state of his particle 3 is the same as that which particle 1 had initially.

Of course, since there are four orthogonal Bell states, there are four equally probable outcomes for Alice's Bell-state measurement. If Alice has obtained another result, the state of Bob's particle is again related to the initial state of particle 1, up to a characteristic unitary transformation. This stems from the fact that a unitary transformation of one of two entangled particles can transform from any Bell state to any other.

Therefore, Alice has to send the result of her Bell-state measurement (i.e. a number between 0 and 3, or, equivalently, two bits of information) via a classical communication channel to Bob. He then can restore the initial quantum state of particle 1 on his particle 3 by performing the correct unitary transformation.

Formally, we describe the initial state of particle 1 by $|\chi\rangle_1 = a|H\rangle_1 + b|V\rangle_1$, and the state of the EPR pair 2 and 3 by $|\Psi^-\rangle_{2,3}$. Then the joint three-photon system is in the product state

$$|\Psi\rangle_{1,2,3} = |\chi\rangle_1 \otimes |\Psi^-\rangle_{2,3}, \tag{3.10}$$

which can be decomposed into

$$|\Psi\rangle_{1,2,3} = \frac{1}{2} \left[-|\Psi^-\rangle_{1,2} (a|H\rangle_3 + b|V\rangle_3) - |\Psi^+\rangle_{1,2} (a|H\rangle_3 - b|V\rangle_3) \right. \tag{3.11}$$
$$\left. + |\Phi^-\rangle_{1,2} (a|V\rangle_3 + b|H\rangle_3) + |\Phi^+\rangle_{1,2} (a|V\rangle_3 - b|H\rangle_3) \right].$$

One easily sees that after observation of particles 1 and 2 in one of the four Bell states, the corresponding unitary transformation enables Bob to transfer the initial state of particle 1 to particle 3.

Some Remarks The principle of quantum teleportation incorporates all the characteristic features of entangled systems, and, in an astounding manner, profits from the obstacles seemingly imposed by quantum mechanics. It should be emphasized that quantum teleportation is well within the concepts of conventional physics and quantum mechanics. Let us briefly discuss a few not infrequently occurring misunderstandings.

First, the no-cloning theorem is not violated. The state of particle 1 can only be restored on particle 3 if the measurement performed by Alice does not give any information about the state! After Alice's Bell-state measurement, particle 1 is in a mixed state, which is absolutely uncorrelated with the initial state of particle 1. Therefore, the particular quantum state which is teleported can be attributed to only one particle at a time, never to two.

Secondly, there is no faster-than-light communication achieved in quantum teleportation. Even if Alice knows, right after her measurement, that Bob's particle is already either in the correct state or in one of the three other possible states, she has to send this information to Bob. The classical information sent to Bob is transmitted, according to the theory of relativity, at the speed of light at maximum. Only after receiving the result and after performing the correct unitary transformation can Bob restore the initial quantum state. If Bob does not know the result of Alice's measurement, his particle is in a mixed state, which is not correlated at all with the initial state. Thus quantum information, the qubit, cannot be transferred faster than classical information.

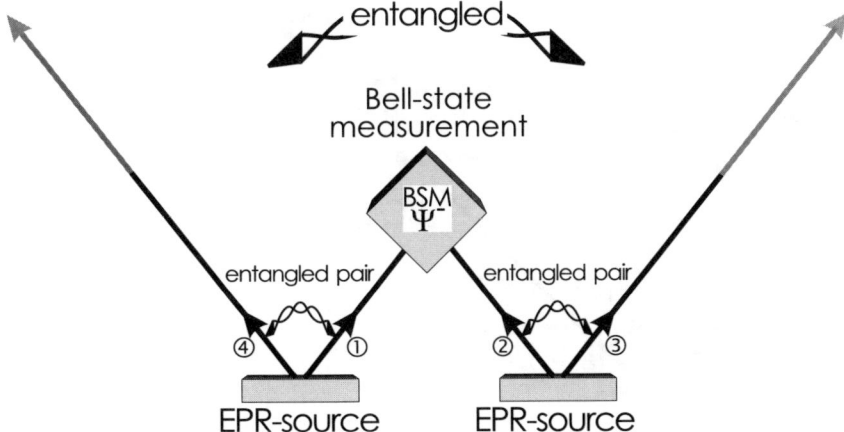

Fig. 3.4. Scheme for entangling particles that have never interacted by the process of entanglement swapping [125]

And, thirdly, there is also no transfer of matter or energy (other than that required for the transmission of classical information). All that makes up a particle are its properties, described by the quantum state. For example, the state of a free neutron defines its momentum and its spin. If one transfers the state onto another neutron, this particle obtains all the properties of the first one; in fact, it becomes the initial particle. We leave it to the science fiction writers to apply the scheme to bigger and bigger objects. Whether or not this idea will help some Captain Kirk to get back to his space ship or not cannot be answered here. Certainly, a lot of other problems need to be solved as well.[5]

It is appropriate to point out some generalizations of the principle of quantum teleportation. It is not necessary that the initial state which is to be teleported is a pure state. In fact it can be any mixed state, or even the undefined state of an entangled particle. This is best demonstrated by entanglement swapping [125]. Here, the particle to be teleported (1) is entangled with yet another one (4) (Fig. 3.4). The state of particle 1 on its own is a mixed state; however, it can be determined by the observation of particle 4. Quantum teleportation allows us to transfer the state of particle 1 onto particle 3. Since quantum teleportation works for any arbitrary quantum state, particle 3 thus becomes entangled with particle 4. Note, that particles 3 and 4 do not come from the same source, nor did they ever interact with each other. Nevertheless, it is possible to entangle them by swapping the entanglement in the process of quantum teleportation.

---

[5] The "technical manuals of *Star Trek*" mention, as a necessary part of their transporter, a "Heisenberg compensator" [124]. Quantum teleportation seems to provide a solution for this marvelous device. However, a lot more is necessary to beam large objects.

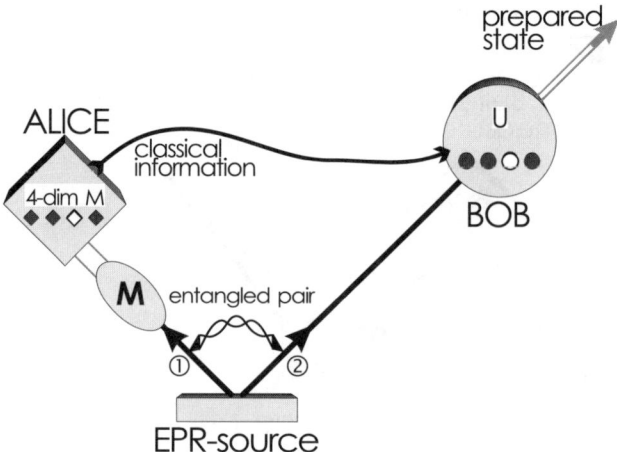

Fig. 3.5. Remote state preparation of Bob's particle 2, by a manipulation (M) of particle 1

Quantum teleportation is not confined to transferring two-state quantum systems. If Alice and Bob share an entangled pair of N-state particles, they can teleport the state of an N-dimensional quantum system [126]. As before, Alice performs projection onto the N^2-dimensional basis of entangled states spanning the product space of particles 1 and 2. The result, one out of N^2 equally probable results, has to be communicated to Bob, who then can again restore the initial state of particle 1 by the corresponding unitary transformation of his particle 3. If Alice and Bob share a pair of particles that are entangled in the original sense of EPR, that is, for continuous variables or ∞-dimensional states, they also can teleport properties such as the position and momentum of particles or the phase and amplitude of electromagnetic fields [11].

A considerable simplification of quantum teleportation, especially in terms of experimental realization, transfers not the quantum state of a particle, but rather the manipulation performed on the entangled particle which is given to Alice [127] (Fig. 3.5). Again, one first distributes an entangled pair to Alice and Bob. But before Alice gets hold of her particle 1 and can perform measurements on it, the state of this particle is manipulated in another degree of freedom. One cannot talk about a two-state system anymore. Rather, particle 1 now is described in a four-dimensional Hilbert space, spanned by the original degree of freedom and the new one. Formally, however, this mimics the two two-state particles given to Alice in the standard quantum teleportation scheme. Consequently, a measurement in the four-dimensional Hilbert space of particle 1, which perfectly erases the quantum information by mixing the two degrees of freedom, is performed. This gives the necessary information for Bob to perform the correct unitary transformation on the particle. That way, the originally mixed state of particle 2 can be turned into a pure state which depends on the manipulation initially performed on particle 1. Using such a scheme, one can remotely prepare particle 3 in any pure quantum state. Thus, it is not necessary to send two real numbers to Bob if one wants him to have a certain, pure quantum state prepared on his particle. If he is provided with one of a pair of entangled particles, Alice simply has to transmit two bits of classical information to Bob.

3.4 The Experimental Prerequisites

Before turning to the fascinating applications of entangled systems, let us review how to produce, how to manipulate and how to measure such quantum systems experimentally. The last decade saw incredible progress in the experimental techniques for handling various quantum systems. However, there are additional challenges when working with entangled systems, especially the careful control of interactions and decoherence of the quantum systems.

In their seminal work Einstein, Podolsky and Rosen considered particles which interacted with each other for a certain time and which thus became entangled and thereafter exhibited the puzzling, nonclassical correlations. The interaction needed to entangle a pair of particles is just the same as von Neumann had in mind when describing the measurement process. Ideally, it couples two quantum systems in such a way that, if the first system is in one of a set of distinguishable (orthogonal) states, the second system will change into a well-defined, corresponding state. Let us look at such a coupling for the simplest case of two two-state systems. As before, the two basis states are denoted as $|0\rangle$ and $|1\rangle$. The coupling is such that if system 1 is in state $|0\rangle_1$, system 2 will remain in its initial state, say $|0\rangle_2$, whereas if system 1 is in state $|1\rangle_1$, system 2 will flip to the orthogonal state, i.e. to $|1\rangle_2$. The nonclassical features arise if system 1 is in a superposition of its basis states. Then, coupling it with the second system results in an entangled state:

$$\begin{aligned} |0\rangle_1 |0\rangle_2 &\longrightarrow |0\rangle_1 |0\rangle_2, \\ |1\rangle_1 |0\rangle_2 &\longrightarrow |1\rangle_1 |1\rangle_2, \\ \frac{1}{\sqrt{2}} \left(|0\rangle_1 + |1\rangle_1 \right) |0\rangle_2 &\longrightarrow \frac{1}{\sqrt{2}} \left(|0\rangle_1 |0\rangle_2 + |1\rangle_1 |1\rangle_2 \right). \end{aligned} \quad (3.12)$$

Although this basic principle of producing entangled states has been known since the very beginning of quantum mechanics, until recently there was no physical system where the necessary coupling could be realized. The progress in cavity QED [36] and ion trap experiments [128] allowed the first observation of entanglement between two atoms or two ions. These experiments are of great importance for the further development of experimental quantum computation. However, for quantum communication one needs to transfer the entangled particles over reasonable distances. Thus photons (with wavelengths in the visible or near infrared) are clearly a better choice. For entangling photons via such a coupling, various methods have been proposed and partially realized [129–131] but still need to be investigated more thoroughly. Fortunately, the process of parametric down-conversion offers an ideal source of entangled photon pairs without the need for strong coupling (see Sect. 3.4.2).

To perform Bell-state analysis, one first has to transform the entangled state into a product state. This is necessary since two particles can be analyzed only if they are measured separately. Otherwise one would need to entangle the two measurement apparatuses, each of which analyzes one of the two particles – clearly an even more challenging task. In principle, a disentangling transformation can be performed by reversing the entangling interaction described above. However, as long as such couplings are not achievable, one has to find replacements. In the following it is shown how two-particle interference can be employed for partial Bell-state analysis (see Sect. 3.4.3). Since the manipulations and unitary transformations have to be performed on only one quantum particle at a time, this does not create new obstacles. These operations are often routine; in the case of light they have been routine for two centuries.

3.4.1 Entangled Photon Pairs

Entanglement between photons cannot be generated by coupling them via an interaction yet. However, there are several emission processes, such as atomic cascade decays or parametric down-conversion, where, owing to the conservation of energy and of linear or angular momentum, the properties of two emitted photons become entangled.

Historically, entanglement between spatially separated quanta was first observed in measurements of the polarization correlation between $\gamma^+\gamma^-$ emissions in positron annihilation [132], soon after Bohm's proposal for observing EPR phenomena in spin-1/2 systems. After Bell's discovery that contradictory predictions between quantum theories can actually be observed, a series of measurements was performed, mostly with polarization-entangled photons from a two-photon cascade emission from calcium [133]. In these experiments, the two photons were in the visible spectrum, and thus could be manipulated and controlled by standard optical techniques. Of course, this is a great advantage compared with the positron annihilation source; however, the two photons are now no longer emitted in opposite directions, since the emitting atom carries away some randomly determined momentum. This makes experimental handling more difficult and also reduces the brightness of the source. The process of parametric down-conversion now offers a new possibility for efficiently generating entangled pairs of photons [111].

When light propagates through an optically nonlinear medium with second-order nonlinearity $\chi^{(2)}$ (only possible in noncentrosymmetric crystals), the

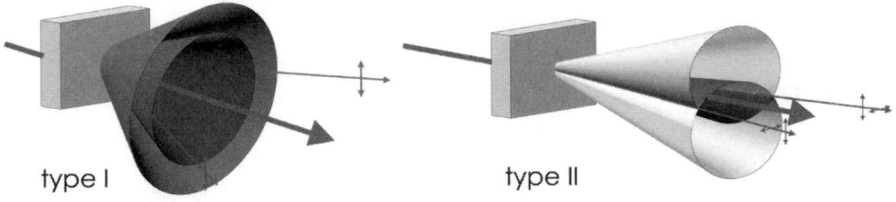

Fig. 3.6. The different relations between the emission directions for type I and type II down-conversion

conversion of a light quantum from the incident pump field into a pair of photons in the "idler" and "signal" modes can occur. In principle, this can be seen as the inverse of the frequency-doubling process in nonlinear optics [134].

As mentioned above, energy and momentum conservation can give rise to entanglement in various degrees of freedom, such as position–momentum and time–energy entanglement. However, the interaction time and volume will determine the sharpness and quality of the correlations observed, which are formally obtained by integration of the interaction Hamiltonian [135]. The interaction time is given by the coherence time τ_c of the UV pump light; the volume is given by the extent and spatial distribution of the pump light in the nonlinear crystal.

The relative orientations of the direction and polarization of the pump beam, and the optic axis of the crystal determine the actual direction of the emission of any given wavelength. We distinguish two possible alignment types (Fig. 3.6): for type I down-conversion, the pump has, for example, the extraordinary polarization and the idler and signal beams both have the ordinary polarization. Different colors are emitted into cones centered on the pump beam.

In type II down-conversion, the pump has the extraordinary polarization and, in order to fulfill the momentum conservation condition inside the crystal (phase-matching), the two down-converted photons have different, for most directions orthogonal, polarizations, offering the possibility of a new source of polarization-entangled photon pairs (Sect. 3.4.2).

One can distinguish two basic ways to observe entanglement. In the first way, by selecting detection events one can chose a subensemble of possible outcomes which exhibits the nonclassical features of entangled states.[6] This additional selection seems to contradict the spirit of EPR–Bell experiments; however, it was shown recently, that, after a detailed analysis of all detection events, the validity of local hidden-variable theories can be tested on the basis

---

[6] For the observation of polarization entanglement, see [136]. For momentum entanglement, see the proposal [137] and the experimental results [138]. Time–energy entanglement was proposed in [139]. Experiments are described in [140].

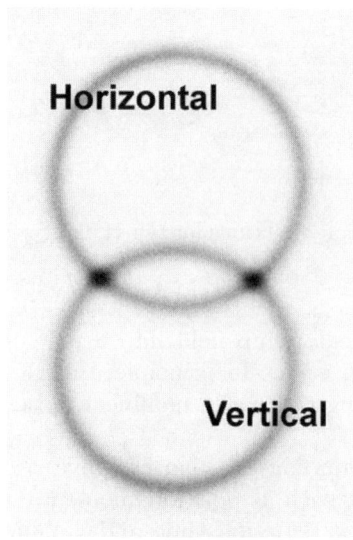

Fig. 3.7. Photons emerging from type II down-conversion. The photons are always emitted with the same wavelength but have orthogonal polarizations. At the intersection points their polarizations are undefined but different, resulting in entanglement

of refined Bell inequalities [141]. Therefore, such sources can also be useful for entanglement-based quantum cryptography [142].

In the second way, true entangled photon pairs can be generated. This is essential for all the other quantum communication schemes, where one cannot use the detection selection method. Several methods to obtain momentum-entangled pairs [143] have been demonstrated experimentally [144], but are extremely difficult to handle experimentally owing to the huge requirements on the stability of the whole setup. Any phase change, i.e. a change in the path lengths by as little as 10 nm, is devastating for the experiment. Also, the recently developed source of time–energy-entangled photon pairs [145] partially shares these problems and, to avoid detection selection, requires fast optical switches. Fortunately, with polarization entanglement as produced by type II parametric down-conversion, the stability requirements are considerably more relaxed.

3.4.2 Polarization-Entangled Pairs from Type II Down-Conversion

In type II down-conversion, the down-converted photons are emitted into two cones, one with the ordinary polarization and the other with the extraordinary polarization. Because of conservation of transverse momentum the photons of each pair must lie on opposite sides of the pump beam. For the proper alignment of the optic axis of the nonlinear crystal, the two cones intersect along two lines (see Fig. 3.7) [111, 146]. Along the two directions ("1" and "2") where the cones overlap, the light can be essentially described by an entangled state:

$$|\Psi\rangle = \left(|H\rangle_1|V\rangle_2 + e^{i\alpha}|V\rangle_1|H\rangle_2\right),\tag{3.13}$$

where the relative phase α arises from the crystal birefringence, and an overall phase shift is omitted. Using an additional birefringent phase shifter (or even by slightly rotating the down-conversion crystal itself), the value of α can be set as desired, e.g. to the value 0 or π. Thus, polarization-entangled states are produced directly out of a single nonlinear crystal (beta barium borate, BBO), with no need for extra beam splitters or mirrors and no requirement to discard detected pairs.

Best of all, by using two extra birefringent elements, one can easily produce any of the four orthogonal Bell states. For example, when starting with the state $|\Psi^+\rangle$, a net phase shift of π and thus a transformation to the state $|\Psi^-\rangle$ may be obtained by rotating a quarter-wave plate in one of the two paths by 90° from the vertical to the horizontal direction. Similarly, a half-wave plate in one path can be used to change a horizontal polarization to vertical and to switch to the states $|\Phi^\pm\rangle$.

The birefringent nature of the down-conversion crystal complicates the actual entangled state produced, since the ordinary and the extraordinary photons have different velocities inside the crystal, and propagate along different directions even though they become parallel and, for short crystals, collinear outside the crystal. The resulting longitudinal and transverse walk-off between the two polarizations in the entangled state is maximal for pairs created near the entrance face of the crystal, which consequently acquire the greatest time delay and relative lateral displacement. Thus the two possible emissions become, in principle, distinguishable by the order in which the detectors would fire or by their spatial location, and no entanglement will be observable. However, the photons are produced coherently along the entire length of the crystal. One can thus completely compensate for the longitudinal walk-off and partially for the transverse walk-off by using two additional crystals, one in each path [147]. By verifying the correlations produced by this source, one can observe strong violations of Bell's inequalities (modulo the typical auxiliary assumptions) within a short measurement time [31].

The experimental setup is shown in Fig. 3.8a. The 351.1 nm pump beam (150 mW) is obtained from a single-mode argon ion laser, followed by a dispersion prism to remove unwanted laser fluorescence (not shown) [111]. Our 3 mm long BBO crystal was nominally cut such that θ^{pm}, the angle between the optic axis and the pump beam, was 49.2°, to allow collinear, degenerate operation when the pump beam is precisely orthogonal to the surface. The optic axis was oriented in the vertical plane, and the entire crystal was tilted (in the plane containing the optic axis, the surface normal and the pump beam) by 0.72°, thus increasing the effective value of θ^{pm} inside the crystal to 49.63°. The two cone overlap directions, selected by irises before the detectors, were consequently separated by 6.0°. Each polarization analyzer consisted of two-channel polarizers (polarizing beam splitters) preceded

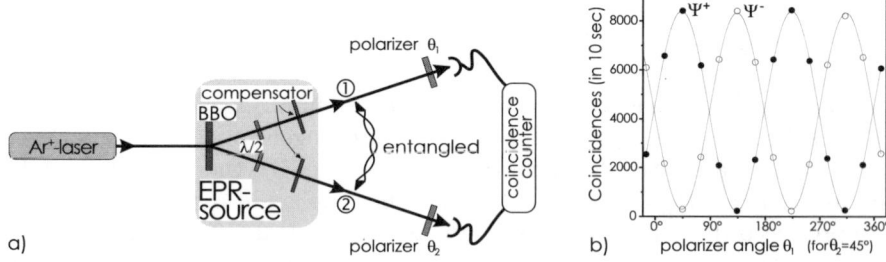

Fig. 3.8. (a) Experimental setup for the observation of entanglement produced by a type II down-conversion source. The additional birefringent crystals are needed to compensate for the birefringent walk-off effects from the first crystal. (b) Coincidence fringes for the Bell states $|\Psi^+\rangle$(•) and $|\Psi^+\rangle$(○) obtained when varying the analyzer angle Θ_1, with Θ_2 set to 45°

by a rotatable half-wave plate. The detectors were cooled silicon avalanche photodiodes operated in the Geiger mode. Coincidence rates $C(\theta_1, \theta_2)$ were recorded as a function of the polarizer settings θ_1 and θ_2.

In this experiment the transverse walk-off (0.3 mm) was small compared with the coherent pump beam width (2 mm), so the associated labeling effect was minimal. However, it was necessary to compensate for the longitudinal walk-off, since the 3.0 mm BBO crystal produced a time delay which was about the same as the coherence time of the detected photons ($\approx$390 fs, determined by interference filters with a width of 5 nm at 702 nm). We used an additional BBO crystal (1.5 mm thick) as a compensator in each of the paths, preceded by a half-wave plate to exchange the roles of the horizontal and vertical polarizations.

Under such conditions, we now obtain routinely a coincidence fringe visibility (as polarizer 2 is rotated, with polarizer 1 fixed at −45°) of more than 97%, for irises with a size of 2 mm at a distance of 1.5 m from the crystal (Fig. 3.8b). The high quality of this source is crucial for the overall performance of our experiments in quantum dense coding [114], quantum cryptography [113] and tests of Bell's inequalities [31]. For the later experiments, the photons were coupled into single-mode fibers, to bridge long distances of the order of 400 m. To achieve a high coupling, the pump beam should be slightly focused into the BBO crystal, to optimally match the microscope objectives used. Since the compensation crystals partially compensate the transverse walk-off, focusing down to 0.2 mm is not crucial. Visibilities of more than 98% have been obtained this way, with an overall collection and detection efficiency of 10%.

Such a source has a number of distinct advantages. It seems to be relatively insensitive to larger collection irises, an important feature in experiments where high count rates are crucial. In addition, owing to its simplicity, the source is much quicker to align than other down-conversion setups and is remarkably stable. One of the reasons is that phase drifts are not detrimental to a polarization-entangled state unless they are birefringent, i.e. polarization-dependent – this is a clear advantage over experiments with momentum-entangled or energy–time-entangled photon pairs. Recently, Kwiat and coworkers tested sandwiched type I crystals and achieved, for thin crystals, a significantly higher relative yield of entangled photon pairs [148]. Also, utilizing cavities to enhance the pump field in the nonlinear crystal can boost the output by a factor of 20 [149]. This gives hope that even more efficient generation of entangled photon pairs will be obtained in the future.

3.4.3 Interferometric Bell-State Analysis

At the heart of Bell-state analysis of a pair of particles is the transformation of an entangled state to an unentangled, product state. The necessary coupling, however, has not been achieved for photons yet. But it turns out that interference of two entangled particles, and thus the photon statistics behind beam splitters depend on the entangled state that the pair is in [112,150, 151].

The Principle Let us discuss first the generic case of two interfering particles. If we have two otherwise indistinguishable particles in different beams and overlap these two beams at a beam splitter, we ask ourselves, what is the probability to find the two particles in different output beams of the beam splitter (Fig. 3.9a). Alternatively we can ask, what is the probability that two detectors, one in each output beam, detect one photon each.

If we performed this experiment with fermions, we would at first naively expect the two fermions to arrive in different output beams. This is suggested by the Pauli principle, which requires that the two particles cannot be in the same quantum state, that is, they cannot exit in the same output beam. Analogously, interference of bosons at a beam splitter will result in the expectation of finding both bosons in one output beam. For a symmetric 50/50 beam splitter, it is fully random whether the two bosons will be detected in the upper or the lower detector, but they will be always detected by the same detector. However, it is important to realize that the statements above are only correct if one disregards the internal degrees of freedom of the interfering particles.

Ultimately, the reason for the different behaviors lies in the different symmetries of the wave functions describing bosonic and fermionic particles [150]. There are four different possibilities for how the two particles could propagate from the input to the output beams of the beam splitter. We obtain one particle in each output if both particles are reflected or both particles are transmitted; we observe both particles at one detector if one particle is transmitted and the other reflected, or vice versa. For the antisymmetric states of fermions, the two possibilities of both particles being transmitted

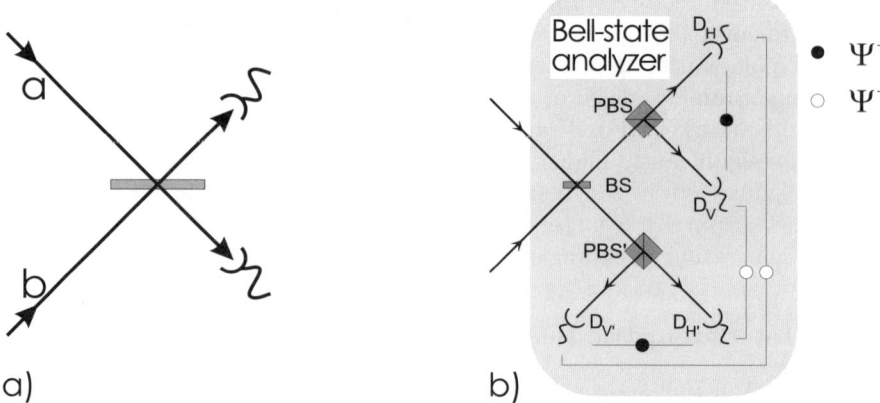

Fig. 3.9. (a) Interference of two particles at a beam splitter. The observation of coincident detection, i.e. detection of one particle at each of the two detectors, is sensitive to the symmetry of the spatial component of the quantum state of the combined system. (b) Bell-state analyzer for identifying the Bell states $|\Psi^+\rangle$ and $|\Psi^-\rangle$ by observing different types of coincidences. The other two Bell states $|\Phi^\pm\rangle$ exhibit the same detection probabilities (both photons are detected by one detector) for this setup and cannot be distinguished

and both being reflected interfere constructively, resulting in firing of each of the two detectors. For the symmetric states of bosons, these two amplitudes interfere destructively, giving no simultaneous detection in different output beams [152]. For photons with identical polarizations, which means for bosons, this interference effect has been known since the experiments by Hong et al. [153],[7] but up to now it has not been observed for fermions yet.

What kinds of interference effects of two photons at a beam splitter are to be expected if we consider also the internal degree of freedom of the photons, i.e. their polarization? In particular, if we interfere two polarization-entangled photons at a beam splitter, the Bell state describes only the internal degree of freedom. Inspection of the four Bell states shows that the state $|\Psi^-\rangle$ is antisymmetric, whereas the other three are symmetric. However, if two particles interfere at a nonpolarizing beam splitter, what matters is only the spatial part of the wave function. The symmetry of the wave function is determined by the requirement that for two photons, the total state has to be symmetric again. We therefore obtain, for the total state of two photons in the antisymmetric Bell state formed from two beams a and b at the beam splitter,

$$|\Psi\rangle = \frac{1}{2}\left(|H\rangle_1|V\rangle_2 - |V\rangle_1|H\rangle_2\right)\left(|a\rangle_1|b\rangle_2 - |b\rangle_1|a\rangle_2\right). \tag{3.14}$$

---

[7] For further experiments and theoretical generalizations, see [154].

This means that, for the state $|\Psi^-\rangle$, we also have an antisymmetric spatial part of the wave function and thus expect a different detection probability, that is, different coincidences between the two detectors, compared with the other three Bell states.

We therefore can discriminate the state $|\Psi^-\rangle$ from all the other states. It is the only one which leads to coincidences between the two detectors in the output beams of the beam splitter. Can we also identify the other Bell states? If two photons are in the state $|\Psi^+\rangle$, they will both propagate in the same output beam but with orthogonal polarizations in the horizontal/vertical (H/V) basis, whereas two photons in the state $|\Phi^+\rangle$ or in the state $|\Phi^-\rangle$, which also both leave the beam splitter in the same output arm, have the same polarization in the H/V basis. Thus we can discriminate between the state $|\Psi^+\rangle$ and the states $|\Phi^\pm\rangle$ by a polarization analysis in the H/V basis and by observing either coincidences between the outputs of a two-channel polarizer or both photons again in only one output (Fig. 3.9b). Note that reorientation of the polarization analysis allows one to separate any other of these three states from the other two, but it is not possible to distinguish between all of them simultaneously [155]. If the photons were entangled in yet another degree of freedom, i.e. they were four-state systems rather than regular qubits, one could also discriminate between the states $|\Phi^+\rangle$ and $|\Phi^-\rangle$ [156]. But up to now, no quantum communication scheme seems to have profited from this fact.

Summarizing, we conclude that two-photon interference can be used to identify two of the four Bell states, with the other two giving the same third detection result. One thus cannot perform complete Bell-state analysis by these interferometric means, but we can identify three different settings in quantum dense coding and, for teleportation, even identification of only one of the Bell states is sufficient to transfer any quantum state from one particle to another, although then only in a quarter of the trials.

Bell-State Analysis of Independent Photons The above description of how to apply two-photon interference for Bell-state analysis can give only some hints about the possible procedures. One intuitively feels that the necessary joint detection of the two photons has to be "in coincidence". But what really are the experimental requirements for the two photons to interfere? The coincidence conditions can be obtained using a more refined analysis that takes the multimode nature of the states involved into account [157].

Interference occurs only if the contributing possibilities for finding one photon in each output are indistinguishable. If the two photons come from different sources or, as is the case in the experiments, from different down-conversion emissions, there might be some timing information, in our case detection of the second photon from each down-conversion, which might render the possibilities distinguishable.

For example, if we detect one photon behind the beam splitter at almost the same time as one of the additional down-conversion photons, we can infer the origin of the photon that is to interfere. However, if the time difference between the detection events of the two interfering photons, that is, the overlap at the beam splitter, is much less than their coherence time, then the detection of any other photon cannot give any additional information about their origin. This ultra-coincidence condition requires the use of narrow filters in order to make the coherence time as long as possible. However, even if we consider using state-of-the-art interference filters that yield a coherence time of about 3 ps, no detectors fast enough exist at present. And an even stronger filtering by Fabry–Perot cavities (to achieve the necessary coherence time of about 500 ps) results in prohibitively low count rates. Only a considerable increase of the number of photon pairs emitted into a narrow wavelength window may allow one to use this technique (e.g. with a subthreshold OPO configuration as demonstrated in [158]).

The best choice, as it turns out, is not to try to detect the two photons simultaneously, but rather to generate them with a time definition much better than their coherence time. Consider two down-conversion processes pumped by pulsed UV beams (using either two crystals or, as is the case in our experiments, one crystal pumped by two passages of a UV beam). Again we attempt to observe interference between two photons, one from each down-conversion process. Then, without any narrow filters in the beams, the tight time correlation of the photons coming from the same down-conversion permits one again to associate simultaneously detected photons with each other. This provides path information and hence prohibits interference.

We now insert filters before (or behind) the beam splitter. With standard filters, and thus also with high enough count rates, one easily achieves coherence times on the order of 1 ps. And it is possible to pump the two down-conversion processes with UV pulses with a duration shorter than 200 fs. Thus it follows that the photons detected behind the beam splitter carry practically no information anymore on the detection times of their twin photons, and, vice versa, detection of those latter photons does not give which-path information, which would destroy the interference.

The "coincidence time" for registering the photons now can be very long; it merely needs to be shorter than the repetition time of the UV pulses, which is on the order of 10 ns for commercially available laser systems. One thus can expect very good visibility of interference and very good precision of the Bell-state analysis.

3.4.4 Manipulation and Detection of Single Photons

For polarization-entangled photons, the unitary transformations transforming between the four Bell states can be performed with standard half-wave and quarter-wave retardation plates.

As mentioned before, in order to have the maximum freedom in setting any of the Bell states, one inserts one half-wave and one quarter-wave plate into the beam. By precompensating the additional quarter-wave shifts with the compensator plates of the EPR source, one obtains at the output of the transformation plates the state $|\Psi^-\rangle$ if both optic axes are aligned along the vertical direction. Rotation of only the quarter-wave plate to the horizontal direction transforms this state to $|\Psi^+\rangle$, and rotation of only the half-wave plate by 45° gives $|\Phi^-\rangle$. Finally, rotating one plate by 90° and the other one by 45° gives $|\Phi^+\rangle$.

For initial experimental realizations of the ideas of quantum communication, such a static polarization manipulation is sufficient. However, for quantum cryptography, and also for practical applications of other schemes, one would like to be able to switch the unitary transformation rapidly to any position. This can be achieved by fast Pockels cells. Depending on the applied voltage, these devices have different indices of refraction for two orthogonal polarization components, and can be used in a similar way to the quartz retardation plates [31].

Detection of the single photons has been performed using silicon avalanche photodiodes operated in the Geiger mode. The diodes used have a detection efficiency of about 40%. Owing to losses in the interference filters and other optical components, the overall detection efficiency of a photon emitted from the source was around 10% in cw experiments; for experiments using a pulsed source, we achieved an efficiency of only about 4%. In many interference experiments, a good definition of the transverse-mode structure of the beams is necessary. An ideal solution for achieving high interference contrast is thus to couple the output arms of a beam splitter into single-mode fibers and connect these fibers to pigtailed avalanche photodiodes. The single-mode fiber acts as a very good spatial filter for the transverse modes and couples the light efficiently to the diodes.

3.5 Quantum Communication Experiments

3.5.1 Quantum Cryptography

In the first experiments [159], the researchers concentrated on the distribution of pairs of entangled photons over large lengths of fibers, rather than on including fast, random switching. In these indoor experiments, where the optical fiber was wound on a fiber drum, one of the photons was chosen to have a wavelength of $\lambda = 1300$ nm, and the other in the near infrared for optimal detection efficiency (here the down-conversion was pumped by a krypton ion laser at 460 nm). Time–energy entanglement was used, with asymmetric interferometers at the observer stations and selection of true coincidences. Such a scheme allows the correlated photons to have a wide frequency distribution, and thus a relatively high intensity, since the visibility

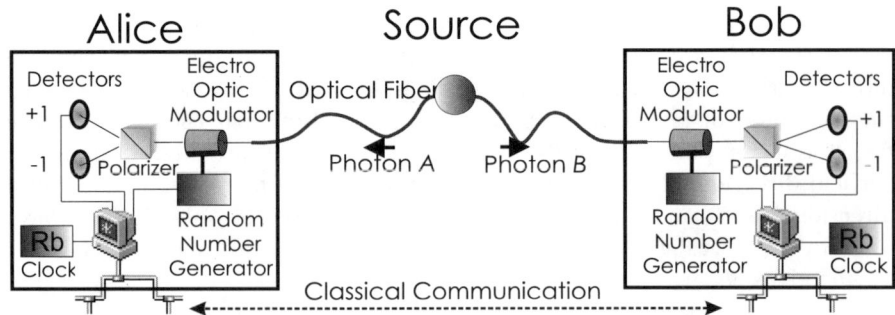

Fig. 3.10. Setup for entanglement-based quantum cryptography. The polarization-entangled photons are transmitted via optical fibers to Alice and Bob, who are separated by 400 m, and both photons are analyzed, detected and registered independently. After a measurement run, the quantum keys are established by Alice and Bob through classical communication over a standard computer network

of the interference effects depends on the monochromaticity of the pump laser light.

In a more recent experiment, both photons were produced with a wavelength around 1300 nm. Here, for the first time, laser diodes ($\lambda = 650$ nm) were used for pumping the down-conversion, in contrast to the expensive laser systems used in other experiments. This allowed the demonstration of nonclassical correlations between two observers separated by more than 10 km in the Geneva area [30]. Standard optical telecommunication fibers connecting offices of the Swiss telecommunication company were used to send the photons to two interferometers, where phase modulation served to set the analysis parameters. The robustness of the source, together with the high degree of quantum entanglement, opens new prospects for this secure communication technique.[8]

The scheme of the first realistic demonstration of entanglement-based key distribution is sketched in Fig 3.10 [113]. The source uses type II parametric down-conversion in BBO, pumped with an argon ion laser working at a wavelength of 351 nm and a power of 350 mW. The photons, with a wavelength of 702 nm, are each coupled into 500 m long optical fibers and transmitted to Alice and Bob, who are separated by 400 m. Alice and Bob both use a Wollaston polarizing beam splitter as a polarization analyzer. We shall associate a detection of parallel polarizations, +1, with the key bit 1 and detection of orthogonal polarizations, −1, with the key bit 0.

Electrooptic modulators in front of the analyzers rapidly switch (<15 ns) the axis of the analyzer between two desired orientations, controlled by quantum random signal generators. These quantum random number generators are based on the quantum mechanical process of splitting a beam of photons

---
[8] For recent experiments on entanglement-based cryptography, see [160].

and have a correlation time of less than 100 ns [161]. The photons are detected by silicon avalanche photodiodes, and time interval analyzers on local personal computers register all detection events as time stamps together with the settings of the analyzers and the detection results.

Quantum key distribution is started by a single light pulse sent from the source to Alice and Bob via a second optical fiber. After a run of about 5 s duration has been completed, Alice and Bob compare their lists of detections to extract the coincidences. In order to record the detection events very accurately, the time bases in Alice's and Bob's time interval analyzers are controlled by two rubidium oscillators. Overall, the system has a measured rate of total coincidences of ~ 1700 per second, and a collection efficiency of each photon path of 5%. All the necessary equipment for the source, Alice and Bob have been proven to operate outside shielded laboratory environments with a very high reliability.

For the realization of entanglement-based quantum cryptography using the Wigner inequality, Alice switches the analyzer randomly between $-30°$ and $0°$, and Bob between $0°$ and $+30°$. After a run, Alice and Bob extract from the coincidences the probabilities $p_{++}(0°, 30°)$, $p_{++}(-30°, 0°)$, and $p_{++}(-30°, 30°)$ for the corresponding analyzer settings. We obtain -0.112 ± 0.014 for the left-hand side of the Wigner inequality (3.7), which is in good agreement with the predictions of quantum mechanics, and the coincidences obtained at the parallel settings, $(0°, 0°)$, can be used as a quantum key. In a typical run, Alice and Bob established 2162 bits of raw quantum key material at a rate of 420 baud, and observed a quantum bit error rate (QBER) of 3.4%. By biasing the frequencies of the analyzer combinations, the production rate of the quantum keys can be increased to about 1700 baud without sacrificing security.

To demonstrate the entanglement-based BB84 scheme, Alice's and Bob's analyzers both switched independently and randomly between $0°$ and $45°$. After a measurement run, Alice and Bob extracted the coincidences measured with parallel analyzers to generate the quantum key. In the experiment, Alice and Bob collected 80 000 bits of quantum key at a rate of 850 baud and observed a quantum bit error rate of 2.5%. To correct the remaining errors and ensure the secrecy of the key, various classical error correction and privacy amplification schemes have been developed. With a very fast and efficient algorithm, a single iteration gives 49 984 bits with a significantly reduced QBER of 0.40% [113].

3.5.2 Quantum Dense Coding

For the first realization of this quantum communication scheme, the experiment consisted of three distinct parts (Fig. 3.11): the EPR source, generating entangled photons in a well-defined state; Alice's station, for encoding the messages by a unitary transformation of her particle; and Bob's Bell-state analyzer, for reading the signal sent by Alice.

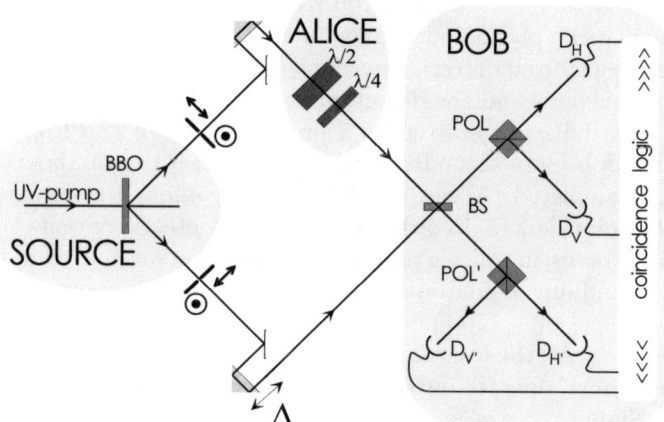

Fig. 3.11. Experimental setup for quantum dense coding. The two entangled photons created by type II down-conversion are distributed to Alice and Bob. Alice sends her photon, after manipulation with birefringent plates, to Bob, who can read the encoded information by interferometric Bell-state analysis. The path length delay Δ is varied to achieve optimal interference

The polarization-entangled photons, with a wavelength of $\lambda = 702$ nm, were, similarly to the quantum cryptography experiment, produced by degenerate noncollinear type II down-conversion in a nonlinear BBO crystal along two distinct emission directions (carefully selected by 2 mm irises, 1.5 m away from the crystal). One beam was directed to Alice's encoding station, the other directly to Bob's Bell-state analyzer. The settings were such that we obtained the entangled state $|\Psi^+\rangle$ behind the compensation crystals (not shown in the figure) and Alice's manipulation unit when the retardation plates were both set to the vertical direction after compensation of birefringence in the BBO crystal.

The beam manipulated in Alice's encoding station was combined with the other beam in Bob's Bell-state analyzer. Bob's analyzer consisted of a single beam splitter followed by two-channel polarizers in each of its outputs, and proper coincidence analysis between four single-photon detectors. In the alignment procedure, optical trombones were employed to equalize the path lengths to well within the coherence length of the down-converted photons ($\Delta\lambda = 100\,\mu$m), in order to observe the two-photon interference.

To characterize the interference observable at Bob's Bell-state analyzer, we varied the path length difference Δ of the two beams with the optical trombone. If the path length difference is larger than the coherence length, no interference occurs and one obtains classical statistics for the coincidence count rates at the detectors. With optimal path length tuning, interference enables one to read the encoded information.

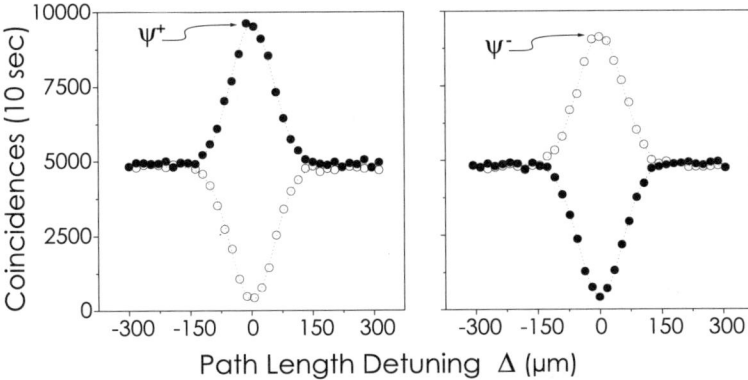

Fig. 3.12. Coincidence rates $\mathcal{C}_{\mathrm{HV}}$ (•) and $\mathcal{C}_{\mathrm{HV'}}$ (o) as functions of the path length difference Δ, when the states $|\Psi^+\rangle$ (*left*) or $|\Psi^-\rangle$ (*right*) are analyzed by Bob's interferometric Bell-state analyzer

Figure 3.12 shows the dependence of the coincidence rates $\mathcal{C}_{\mathrm{HV}}$ (•) and $\mathcal{C}_{\mathrm{HV'}}$ (o) on the path length difference, when either the state $|\Psi^+\rangle$ (left) or the state $|\Psi^-\rangle$ has been sent to the Bell-state analyzer (the rates $\mathcal{C}_{\mathrm{H'V'}}$ and $\mathcal{C}_{\mathrm{H'V}}$ display analogous behavior; we use the notation $\mathcal{C}_{\mathrm{AB}}$ for the coincidence rate between the detectors $\mathrm{D_A}$ and $\mathrm{D_B}$). For perfect path length tuning, $\mathcal{C}_{\mathrm{HV}}$ reaches its maximum for $|\Psi^+\rangle$ (left) and vanishes (apart from noise) for $|\Psi^-\rangle$ (right). $\mathcal{C}_{\mathrm{HV'}}$ displays the opposite dependence and clearly signifies $|\Psi^-\rangle$. The results of these measurements imply that if both photons are detected, we can identify the state $|\Psi^+\rangle$ with a reliability of 95%, and 93% for the state $|\Psi^-\rangle$.

The performance of the dense-coding transmission is influenced not only by the quality of the alignment procedure, but also by the quality of the states sent by Alice. In order to evaluate the latter, the beam splitter was translated out of the beams. Then an Einstein–Podolsky–Rosen–Bell-type correlation measurement analyzed the degree of entanglement of the source, as well as the quality of Alice's transformations. The correlations were only 1–2% higher than the visibilities with the beam splitter in place, which means that the quality of this experiment is limited more by the quality of the entanglement of the two beams than by that of the interference achieved.

When using Si avalanche diodes in the Geiger mode for single-photon detection, a modification of the Bell-state analyzer is necessary, since then, for the states $|\Phi^\pm\rangle$, one has to register the two photons leaving the Bell-state analyzer via a coincidence detection. One possibility is to avoid interference at all for these states by introducing polarization-dependent delays before Bob's beam splitter. Another approach is to split the incoming two-photon state at an additional beam splitter and to detect it (with 50% likelihood) by a coincidence count between detectors in each output (inset of Fig. 3.13). For the purpose of a proof-of-principle demonstration, we put such a con-

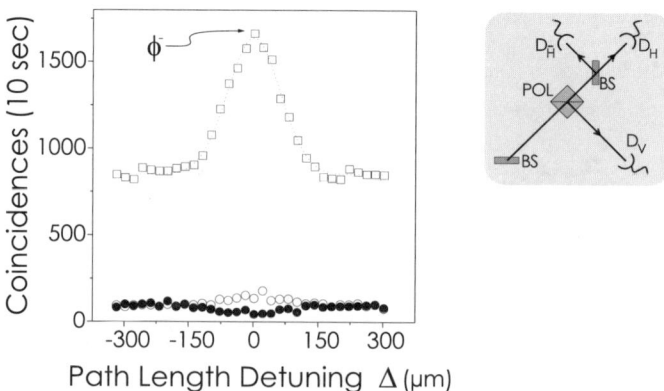

Fig. 3.13. Coincidence rates as functions of the path length difference Δ. Because of the nature of the Si avalanche photodiodes, the extension shown in the *inset* is necessary for identifying two-photon states in one output

figuration in place of detector D_H only. Figure 3.13 shows the increase of the coincidence rate $C_{\overline{HH}}$ ($\square$) for zero path length difference, with the other rates at the background level, when Alice sends the state $|\Phi^-\rangle$. Since we can now distinguish the three different messages, the stage is set for the quantum dense-coding transmission. Figure 3.14 shows the various coincidence rates (normalized to the corresponding maximum rate of the transmitted state), when the ASCII codes of "KM°" (i.e. codes 75, 77, 179) were sent in only 15 trits instead of 24 classical bits.

From this measurement, one can also obtain a signal-to-noise ratio by comparing the rates signifying the actual state with the sum of the two other rates registered. The ratios for the transmission of the three states varied

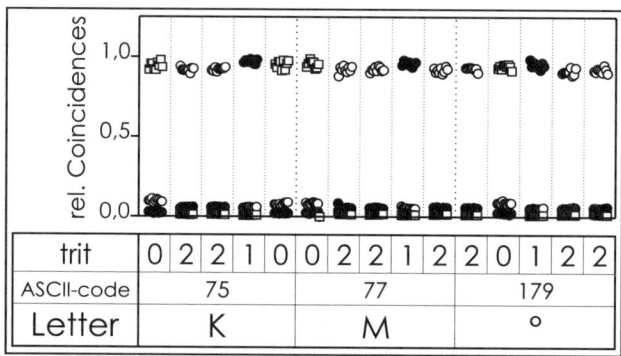

Fig. 3.14. "1.58 bits per photon" quantum dense coding: the ASCII codes for the letters "KM°" (i.e. 75, 77, 179) are encoded in 15 trits instead of the 24 bits usually necessary. The data for each type of encoded state are normalized to the maximum coincidence rate for that state

owing to the different visibilities of the corresponding interferences and were about 14% and 9%. The signal-to-noise ratio achieved results in an actual channel capacity of 1.13 bits per transmitted (and detected) two-state photon and thus clearly exceeds the channel capacity of 1 bit achievable with noise-free classical communication.

3.5.3 Quantum Teleportation of Arbitrary Qubit States

In this experiment, polarization-entangled photons were produced again by type II down-conversion in a nonlinear BBO crystal (see Fig. 3.15), but here the UV beam was pulsed to obtain a high time definition of the creation of the pairs (the pulses had a duration of about 200 fs and $\lambda = 394$ nm). The entangled pair of photons 2 and 3 is produced in the first passage of the UV pulse through the nonlinear crystal, and the pair 1 and 4 after the pulse has been reflected at a mirror back through the crystal. Mirrors and beam splitters (BS) are used to steer and to overlap the light beams. Polarizers (Pol) and polarizing beam splitters (PBS), together with birefringent retardation plates ($\lambda/2$), prepare and analyze the polarization of the photons. All single-photon detectors indicated in the figure (silicon avalanche photodiodes operated in the Geiger mode) are equipped with narrow-band interference filters; the detectors of Alice's Bell-state analyzer are equipped with additional single-mode fiber couplers for spatial filtering.

For the first demonstration of quantum teleportation [10, 11], we prepared particle 1 in various nonorthogonal polarization states using a polarizer and a quarter-wave plate (not shown). Behind Bob's "receiver", polarization analysis was performed to prove the dependence of the polarization of photon 3 on the polarization of photon 1. (In this case we used the registration of photon 4 only to define the time of appearance of photon 1.)

The first task now is to prove that no information about the state of photon 1 is revealed during the Bell-state measurement of Alice. Figure 3.16 shows the coincidence rate between detectors f1 and f2 when the overlap of photons 1 and 2 at the beam splitter was varied (for this, we changed the position of the mirror reflecting the pump beam back into the crystal). The characteristic interference effect, a reduction of the coincidence rate, occurs only around zero delay. Outside this region, which is on the order of the coherence length of the detected photons, no reduction occurs, and the two photons are detected in coincidence with 50% probability. Within the statistics, there is no difference between the two data sets, although particle 1 was prepared in two mutually orthogonal states ($+45°$ and $-45°$). Obviously, Alice has no means to determine which of the two states particle 1 was in after the projection into the Bell-state basis.

Figure 3.17 shows the polarization of photon 3 after the teleportation protocol has been performed, again as the delay between photons 1 and 2 is varied. When interference occurs at the beam splitter, i.e. around zero delay, the polarization of photon 3 is given by the settings for photon 1. The two

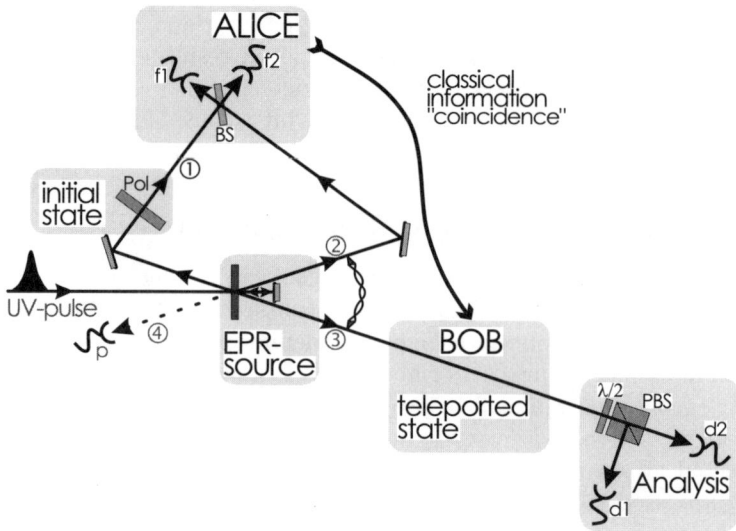

Fig. 3.15. Experimental setup for quantum teleportation. A UV pulse passing through a nonlinear crystal creates an entangled pair of photons 2 and 3 in the state $|\Psi^-\rangle$, which is distributed to Alice and Bob. During its second passage through the crystal, after retroreflection the UV pulse creates another pair of photons, of which one is prepared in the initial state to be teleported (photon 1), and the other one (4) serves as a trigger indicating that a photon to be teleported is on its way. Alice then looks for coincidences behind her beam splitter, where the initial photon and one of the ancillaries are superposed. Bob, after receiving the classical information that Alice has obtained a coincidence count identifying the $|\Psi^-\rangle$ Bell state, knows that his photon 3 is in the initial state of photon 1, which then can be verified using polarization analysis

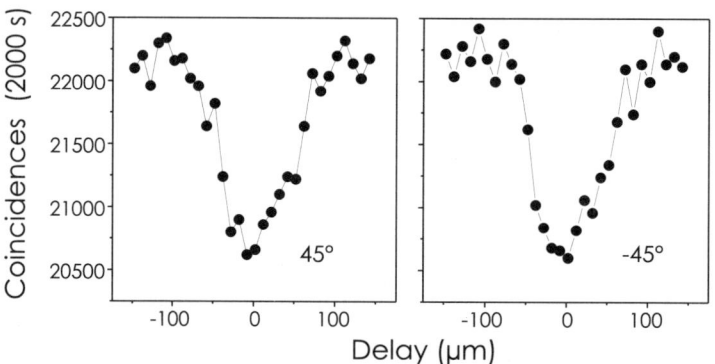

Fig. 3.16. Coincidence rate between the two detectors of Alice's Bell-state analyzer as a function of the delay between the two photons 1 and 2. The data for the $+45°$ and $-45°$ polarizations of photon 1 are equal within the statistics, which shows that no information about the state of photon 1 is revealed to Alice

graphs show the results obtained when the initial polarization of photon 1 was set either to 45° or to vertical polarization and then the polarization of photon 3 along the corresponding direction was analyzed. The reduction in the polarization to about 65% is due to the limited degree of entanglement between photons 2 and 3 (85%), and to the reduced contrast of the interference at the beam splitter as a consequence of the relatively short coherence time of the detected photons. Of course, better beam definition by narrow pinholes and more stringent filtering could improve this value. However, this would cause further, unacceptable loss in the fourfold coincidence rates. Each of the polarization data points shown was obtained from about 100 four-fold coincidence counts in 4000 s.

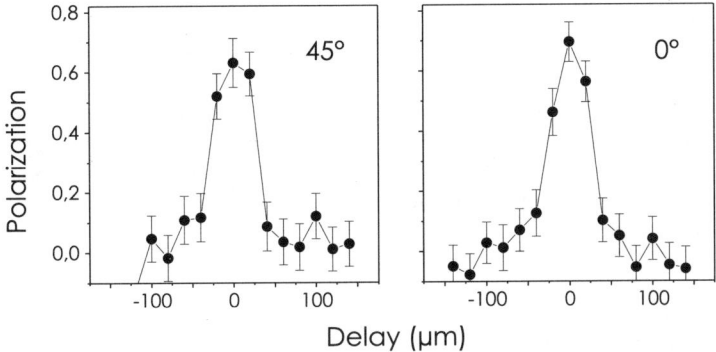

Fig. 3.17. Polarization of photon 3 after teleportation, compared with the polarization initially prepared on photon 1. The analyzer testing the quality of the teleportation performed by Alice and Bob was oriented parallel to the initial polarization

These measurements and also runs with the initial polarization along other directions demonstrate the ability to teleport the polarization of any pure state. Of course, since the directions used are mutually nonorthogonal, one can infer that the scheme works for any arbitrary quantum state. However, there is a much more direct way to experimentally demonstrate the full power of quantum teleportation.

One way to demonstrate that any arbitrary quantum state can be transferred is to use the fact that we can also obtain entanglement between photons 1 and 4 (Fig. 3.18). After the polarizer was removed from arm 1 and put into arm 4, the state of 1 was not defined anymore, but still could be teleported to photon 3; this was demonstrated by showing that now the entanglement had been swapped to photons 3 and 4.

The state of photon 1 (Fig. 3.18), which is part of an entangled pair (photons 1 and 4), is fully undetermined and is formally described by a mixed state. If one can teleport this state to another photon, i.e. to Bob's photon 3, we expect to find this photon in a mixed state, that means it is unpolarized.

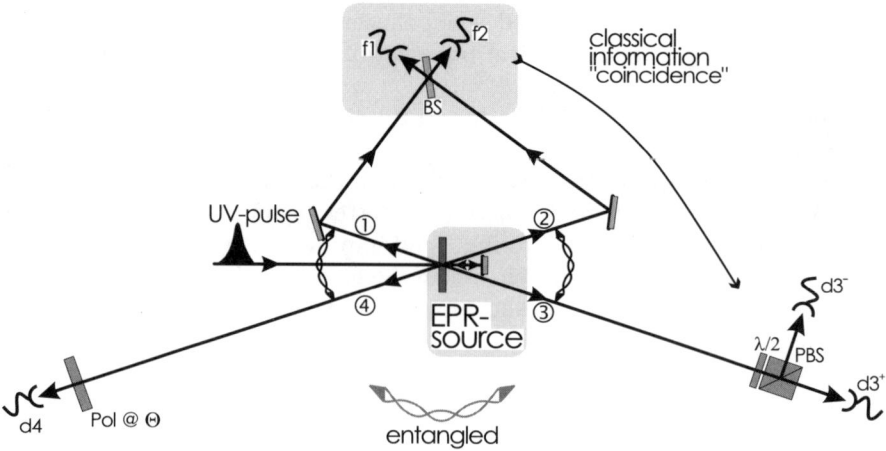

Fig. 3.18. Experimental setup that demonstrates teleportation of arbitrary quantum states: by teleporting the as yet undefined state of photon 1 to photon 3, one is able to swap the entanglement, initially between particles 1 and 4 and between particles 2 and 3, to the newly entangled pair of particles 3 and 4 by projecting 1 and 2 into an entangled pair

Now, since Bob's photon was originally also part of an entangled pair (photons 2 and 3), it was unpolarized anyway. One might conclude that here we did not achieve anything. However, if one determines not only the polarization of photon 3 but the correlations between photons 3 and 4, one finds that now these two photons, which have been produced independently by different processes, are entangled [115].

Figure 3.19 verifies the entanglement between photons 3 and 4, conditioned on coincidence detection of photons 2 and 3. Varying the angle Θ of the polarizer in arm 4 causes a sinusoidal variation of the count rate, here with the analyzer of photon 3 set to $\pm 45°$. This shows that we did not teleport just a mixed state, but actually the as yet undetermined state of the entangled photon.

These experiments present the first demonstration of quantum teleportation, that is, the transfer of a qubit from one two-state particle to another. In the meantime, further steps have been achieved, in particular the remote preparation of the state of Bob's photon (sometimes also called "teleportation") [10] and, especially important, the teleportation of the state of an electro-magnetic field [11]. The latter is the first example of teleportation of continuous variables based on the original EPR entanglement. The first experiment demonstrated the feasibility of transfer of fluctuations of a coherent state from one light beam to another. Although the experiment was limited to a narrow bandwidth of 100 kHz, this was only a technical limitation due to the detection electronics, the modulators and the bandwidth of the source

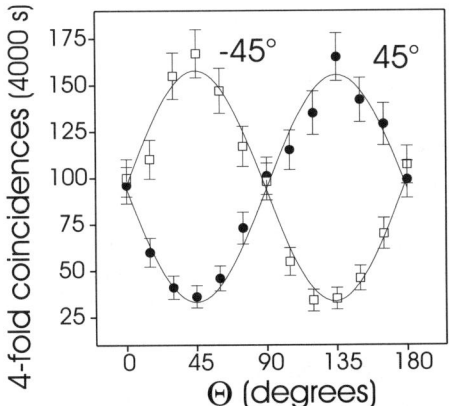

Fig. 3.19. Verification of the entanglement between photons 3 and 4. The sinusoidal dependence of the fourfold coincidence rate on the orientation Θ of the polarizer in arm 4 for $\pm 45°$ polarization analysis of photon 3 demonstrates the possibility to teleport any arbitrary quantum state

of EPR-entangled light beams. In principle, it soon should be possible also to transfer nonclassical states of light, such as squeezed light or number states.

3.6 Outlook

Quantum communication with entangled photons has shown its power and its fascinating features. Our experiments, where realistic entanglement-based quantum cryptography has been performed, where the capacity of communication channels has been increased beyond classical limits and where the polarization state of a photon has been transferred to another one by means of quantum teleportation, are only the first steps towards the exploitation of new resources for communication and information processing.

Quantum communication can offer a wealth of further possibilities, especially when combined with simple quantum logic circuitry. Quantum computers have to operate on large numbers of qubits to really demonstrate their power. But quantum communication schemes already profit from combining only a few qubits and entangled systems. Quantum logic operations with several particles are already useful in examples of the quantum coding theorem [69], but have shown their importance particularly in the proposals for entanglement purification [119]. Any realistic transmission of quantum states will suffer from noise and decoherence along the line. If one wants to distribute entangled pairs of particles to, say, Alice and Bob, the entanglement between the received particles will be considerably degraded, which would prevent successful quantum teleportation, for example. If Alice and Bob now combine the particles of several such noisy pairs on each side by quantum logic operations, they can improve the quality of entanglement by the proposed "distillation" process.

These ideas are closely related to quantum error correction for quantum computers and were recently implemented in a proposal for efficient distribution of entanglement via so-called quantum repeaters [162]. One day, such systems might form the core of quantum networks [163] allowing quantum communication and even computation over large distances. Of course, one should always keep in mind the obstacles put in the way by the decoherence of quantum states [164]. However, quantum communication schemes should be significantly more stable owing to the much lower number of quantum systems involved.

Once entangled particles have been distributed, various quantum communication protocols could be implemented. Besides those described in the preceding sections, there are some recent proposals that give a new twist to quantum information processing. Quantum gambling [165] and quantum games [166], e.g. a "quantized" version of the prisoner dilemma, bring the field of game theory to the quantum world and demonstrate new strategies in well-known classical games. But the new ideas and thoughts might also be quite useful for other types of communication problems. For example, the quantum version of "Chinese whispers" [167] can be also seen as a special type of error correction scheme. Errors in the classical communication, the whispering, can be more efficiently corrected if the sender and receiver have been provided with entangled pairs of particles.

New possibilities arise if entangled triples of particles are used. For certain tasks, the communication between three or more parties becomes less complex, and thus more efficient, if the parties share the entanglement initially [168], and also schemes for quantum cloning [169] of the state of a qubit become feasible with entangled triples.

Now that significant improvements of down-conversion sources [145, 148, 149] and the first observation of three-particle entanglement [38, 170] have been achieved, the realization with entangled triples of those schemes that have previously used entangled pairs is within the reach of future experiments. For realizing entanglement purification and similar schemes, the experiments immediately become much more complex. It first has to be seen what methods can be used to perform quantum logic operations with photons, and also what types of photon sources should be used then. However, the combined progress in the form of improving experimental techniques and of better understanding of the principles of quantum information theory makes the more complicated schemes feasible. Most likely, there will also be novel schemes for quantum communication using higher numbers of qubits and/or even more complex types of entanglement.

Quantum cryptography was the first quantum communication method to literally leave the shielded environment of quantum physics laboratories and to become a promising candidate for commercial exploitation. We expect that the future will show an enormous potential for and benefit from the use of other quantum communication methods, such as the distribution of entanglement over large distances and the transfer of quantum information in the process of quantum teleportation.

4 Quantum Algorithms: Applicable Algebra and Quantum Physics

Thomas Beth and Martin Rötteler

4.1 Introduction

Classical computer science relies on the concept of Turing machines as a unifying model of universal computation. According to the modern Church–Turing Thesis, this concept is interpreted in the form that every physically reasonable model of computation can be *efficiently* simulated on a probabilistic Turing machine. Recently this understanding, which was taken for granted for a long time, has required a severe reorientation because of the emergence of new computers that do *not* rely on classical physics but, rather, use effects predicted by quantum mechanics.

It has been realized that, by using the principles of quantum mechanics, there are problems for which a putative quantum computer could outperform any classical computer. Quantum algorithms benefit from the application of the superposition principle to the internal states of the quantum computer, which are considered to be states in a (finite-dimensional) Hilbert space. As a result, these algorithms lead to a new theory of computation and might be of central importance to physics and computer science. Striking examples of quantum algorithms are Shor's factoring algorithm, Grover's search algorithm and algorithms for quantum error-correcting codes, all of which will be part of this contribution.

We shall introduce the complexity model of quantum gates, which are most familiar to researchers in the field of quantum computing, and shall give many examples of the usefulness and conciseness of this formalism. Quantum circuits provide a computational model equivalent to quantum Turing machines. This means that, very much like the situation in classical computing, there are several ways of describing computations by appropriate theoretical models.

Amongst such quantum circuits, quantum signal transforms form basic primitives in the treatment of controlled quantum systems. A surprising and important result, in view of the algorithms of Shor, is the fact that it is possible to compute a Fourier transform (of size 2^n) on a quantum computer by means of a quantum circuit which requires only $O(n^2)$ basic operations. This is a substantial speedup compared to the classical case, where the fast Fourier transform [171] yields an algorithm that requires $O(n2^n)$ arithmetic operations. Applications of such Fourier transforms to finite abelian groups arise in the algorithms of Simon and Shor. We shall present these algorithms and the underlying principle leading to their surprisingly fast solution on a quantum computer.

As already mentioned, one of the basic results is that in the complexity model of quantum circuits, the Fourier transform can be realized with an exponential speedup compared with the classical case. However, in the quantum regime the only way to extract information from a system is to make measurements and thereby project out nearly all aspects of the whole system. Thus, the art and science of designing quantum algorithms lies in the ability to obtain enough information from measurements, i.e. to choose the right bases from which relevant information can be read off. On the basis of the example of the Fourier observable, which represents the most important case of such a base change, we explain the underlying principle by means of the so-called hidden-subgroup algorithms and present an analysis of sampling in the Fourier basis with respect to the appropriate groups.

We then show how recent results in the theory of signal processing (for a classical computer) can be applied to obtain fast quantum algorithms for various discrete signal transforms, including Fourier transforms for nonabelian groups. Finally, we give a brief introduction to the theory of (quantum) error-correcting codes and their algorithmic implementation.

4.2 Architectures and Machine Models

The definition of an architecture and a machine model, on which the computations are considered to be carried out, is indispensable if one is to have a common computational model for which algorithms can be devised.

Each reasonable model of computation should give us the possibility of performing arbitrary operations, up to a desired accuracy, on the system by execution of elementary operations. By counting the elementary operations necessary to complete a given task, we arrive at complexity models. Finally, if different approaches defining universal computational models are possible, it is desirable to show the equivalence of these models, in the sense that they can simulate each other with a slowdown that is polynomial in the size of the input. In the case of quantum computing, we give two models for universal quantum computation, namely quantum networks in Sect. 4.2.1 and quantum Turing machines in Sect. 4.2.6. We shall put more emphasis on gates and networks, relying on the result that these two models are equivalent in the sense described.

One remark concerning the architecture is in order: we restrict ourselves to the case of operational spaces with a dimension that is a power of two, which are called *qubit architectures*. These systems incorporate the features necessary to do quantum computing, i.e. superposition of an exponentially growing number of states, interference between computational paths and entanglement between quantum registers.

Fig. 4.1. Elementary quantum gates

It is possible to perform an embedding of an arbitrary finite-dimensional operational space into a qubit architecture (see Sect. 4.2.3); however, this reduction involves a suitable encoding of the states of the system into the basis states of the qubit architecture, and hence genuine properties of the system might be lost by this procedure.

4.2.1 Quantum Networks

The state of a quantum computer is given by a normalized vector in a Hilbert space $\mathcal{H}_{2^n}$ of dimension 2^n, which is endowed with a natural tensor structure $\mathcal{H}_{2^n} \cong \mathcal{C}^2 \otimes \ldots \otimes \mathcal{C}^2$ (n factors). The standard basis for this Hilbert space is the set $\{|x\rangle : x \in \mathbf{Z}_2^n\}$ of binary strings of length n. Restricting the computational space to Hilbert spaces of this particular form is motivated by the idea of a quantum register consisting of n quantum bits. A quantum bit, also called a *qubit*, is a state corresponding to one tensor component of $\mathcal{H}_{2^n}$ and has the form

$$|\varphi\rangle = \alpha|0\rangle + \beta|1\rangle \, , \quad \alpha, \beta \in \mathcal{C} \, , \quad |\alpha|^2 + |\beta|^2 = 1 \, .$$

The possible operations that this computer can perform are the elements of the unitary group $\mathcal{U}(2^n)$. To study the complexity of performing unitary operations on n-qubit quantum systems, we introduce the following two types of computational primitives: *local unitary operations* on a qubit i are matrices of the form $U^{(i)} = \mathbf{1}_{2^{i-1}} \otimes U \otimes \mathbf{1}_{2^{n-i}}$, where U is an element of the unitary group $\mathcal{U}(2)$ of 2×2 matrices and $\mathbf{1}_N$ denotes the identity matrix of size N. Furthermore, we need operations which affect two qubits at a time, the most prominent of which is a so-called *controlled NOT gate* (also called a *measurement gate*) between the qubits j (control) and i (target), denoted by $\text{CNOT}^{(i,j)}$. On the basis vectors $|x_n, \ldots, x_1\rangle$ of $\mathcal{H}_{2^n}$, the operation $\text{CNOT}^{(i,j)}$ is defined by

$$|x_n, \ldots, x_{i+1}, x_i, x_{i-1}, \ldots, x_1\rangle \mapsto |x_n, \ldots, x_{i+1}, x_i \oplus x_j, x_{i-1}, \ldots, x_1\rangle \, ,$$

where the addition $\oplus$ is performed in $\mathbf{Z}_2$.

In graphical notation, using quantum wires, these transformations are written as shown in Fig. 4.1. Lines correspond to qubits, unaffected qubits are omitted and a dot • sitting on a wire denotes a control bit. Note that we draw the qubits according to their significance, starting with the most significant qubit on top. Quantum circuits are always read from left to right.

The two types of gates shown in Fig. 4.1 suffice to generate all unitary transformations, i.e. they form a universal set of gates. This is the content of the following theorem [172].

Theorem 4.1. $\mathcal{G}_1 := \{U^{(i)}, \text{CNOT}^{(i,j)} \mid U \in \mathcal{U}(2),\ i,j \in \{1,\ldots,n\},\ i \neq j\}$ is a generating set for the unitary group $\mathcal{U}(2^n)$.

This means that for each $U \in \mathcal{U}(2^n)$ there is a word $w_1 w_2 \ldots w_k$ (where $w_i \in \mathcal{G}_1$ for $i = 1,\ldots,k$ is an elementary gate) such that U factorizes as $U = w_1 w_2 \ldots w_k$. On the basis of theorem 4.1, we now define a complexity measure for unitary operations on qubit architectures.

Definition 4.1. Let $U \in \mathcal{U}(2^n)$ be a given unitary transformation. Then $\kappa(U)$ is defined as the minimal number k of operations in $\mathcal{G}_1$ necessary to write $U = w_1 w_2 \ldots w_k$ as a sequence of elementary gates.

For the complexity measure κ, the following holds: $\kappa(A \otimes B) \leq \kappa(A) + \kappa(B)$ for all $A \in \mathcal{U}(2^{n_1})$ and $B \in \mathcal{U}(2^{n_2})$, because tensor products are free of cost in a computational model based on quantum mechanical principles. Also, by concatenation of operations, we obtain $\kappa(A \cdot B) \leq \kappa(A) + \kappa(B)$ for $A, B \in \mathcal{U}(2^n)$. Note that whereas in the usual linear complexity measure L_c [173, 174] permutation matrices are free (i.e. $L_c(\pi) = 0$, for all $\pi \in S_n$), we have to take them into account when using the complexity measure κ. Instead of the universal set of gates $\mathcal{G}_1$ we can, alternatively, use the set $\mathcal{G}_2 := \{U^{(i,j)} : U \in \mathcal{U}(4),\ i,j \in \{1,\ldots,n\},\ i \neq j\}$ of all two-bit gates, changing the value of κ by only a constant.

Whereas the complexity measure κ is used in cases where we want to implement a given unitary operation U *exactly* in terms of the generating sets $\mathcal{G}_1$ and $\mathcal{G}_2$, it is also expedient to consider unitary *approximations* by quantum networks. By this we mean a sequence of operations $w_1, \ldots, w_n$ which approximates U up to a given ϵ, i.e. such that $\|U - \prod_{i=1}^{n} w_i\| < \epsilon$, where $\|\cdot\|$ denotes the spectral norm.[1] We denote the corresponding complexity measure by κ_ϵ.

Remark 4.1. The following facts concerning approximation by elementary gates are known:

- There are two-bit gates which are universal [175, 176], i.e. there exists a unitary transformation $A \in \mathcal{U}(4)$ with respect to which it is possible to approximate any given U up to $\epsilon > 0$ by a sequence of applications of A to two tensor components of $\mathcal{H}_{2^n}$ only: $\|U - \prod_{i=1}^{k} A^{(m_i, n_i)}\| < \epsilon$. Even though the much stronger statement of universality of a *generic* two-bit gate is known to be true, it is hard to prove the universality for a *given* two-bit gate [176, 177].

---
[1] Recall that the spectral norm of a matrix $A \in \mathcal{C}^{n \times n}$ is given by $\max_{\lambda \in \text{Spec}(A)} |\lambda|$.

- Small generating sets are known; for instance, we can choose

$$\mathcal{G}_3 := \left\{ \frac{1}{\sqrt{2}} \begin{pmatrix} 1 & 1 \\ 1 & -1 \end{pmatrix}, \begin{pmatrix} 1 & 0 \\ 0 & (1+i)\sqrt{2} \end{pmatrix}, \text{CNOT}^{(k,l)}, k \neq l \right\},$$

where the first two operations generate a dense subgroup in $\mathcal{U}(2)$. The Hadamard transformation, which is part of this generating set, is denoted by

$$H_2 := \frac{1}{\sqrt{2}} \begin{pmatrix} 1 & 1 \\ 1 & -1 \end{pmatrix}$$

and is an example of a Fourier transformation on the abelian group $\mathbf{Z}_2$ (see Sect. 4.4).
- Knill has obtained a general upper bound $O(n4^n)$ for the approximation of unitary matrices using a counting argument [178, 179].
- We cite the following approximation result from Sect. 4.2. of [180]: Fix a number n of qubits and suppose that $\langle X_1, \ldots, X_r \rangle = \mathcal{SU}(2^n)$, i.e. $X_1, \ldots, X_r$ generate a dense subgroup in the special unitary group $\mathcal{SU}(2^n)$. Then it is possible to approximate a given matrix $U \in \mathcal{SU}(2^n)$ with given accuracy $\epsilon > 0$ by a product of length $O\{\text{poly}[\log(1/\epsilon)]\}$, where the factors belong to the set $\{X_1, \ldots, X_r, X_1^{-1}, \ldots, X_r^{-1}\}$. Furthermore, this approximation is constructive and efficient, since there is an algorithm with running time $O\{\text{poly}[\log(1/\epsilon)]\}$ which computes the approximating product. However, we remind the reader that this holds only for a fixed value of n; the constant hidden in the O-calculus grows exponentially with n (see Theorem 4.8 of [180]).

From now on, we put the main emphasis on the model for realizing unitary transformations exactly and on the associated complexity measure κ. In general, only exponential upper bounds for the minimal length occuring in factorizations are known. However, there are many interesting classes of unitary matrices in $\mathcal{U}(2^n)$ that lead to only a polylogarithmic word length, which means that the length of a minimal factorization grows asymptotically like $O[p(n)]$, where p is a polynomial.

In the following we give some examples of transformations, their factorization into elementary gates and their graphical representation in terms of quantum gate arrays. The operations considered in these examples admit short factorizations and will be useful in the subsequent parts of this chapter.

Example 4.1 (Permutation of Qubits). The symmetric group S_n is embedded in $\mathcal{U}(2^n)$ by the natural operation of S_n on the tensor components (qubits). Let $\tau \in S_n$ and let Π_τ be the corresponding permutation matrix on 2^n points. Then $\kappa(\Pi_\tau) = O(n)$: to prove this, we first note that each element $\sigma \in S_n$ can be written as a product $\sigma = \tau_1 \cdot \tau_2$ of two involutions τ_1 and $\tau_2 \in S_n$, i.e. $\tau_1^2 = \tau_2^2 = id$. To see that it is always possible to find a suitable

τ_1 and τ_2, we can, when considering the decomposition of σ into disjoint cycles, restrict ourselves to the case of an n-cycle. Now the decomposition follows immediately from the fact that there is a dihedral group of size $2n$ containing σ as the canonical n-cycle and that this rotation is the product of two reflections.

The unitary transformation corresponding to Π_τ, where $\tau \in S_n$ is an involution, can be realized by swappings of quantum wires, which, in turn, can be performed efficiently and in parallel. To swap two quantum wires we can use the well-known identity $\Pi_{(1,2)} = \text{CNOT}^{(1,2)} \cdot \text{CNOT}^{(2,1)} \cdot \text{CNOT}^{(1,2)}$, yielding a circuit of depth three. Writing an arbitrary permutation Π_τ of the qubits as a product of two involutions, we therefore obtain a realization by a circuit of depth six at most (see also [181]).

As an example, the permutation $(1,3,2)$ of the qubits (which corresponds to the permutation $(1,4,2)(3,5,6)$ on the register) is factored as $(1,3,2) = (1,2)(2,3)$ (see Fig. 4.2).

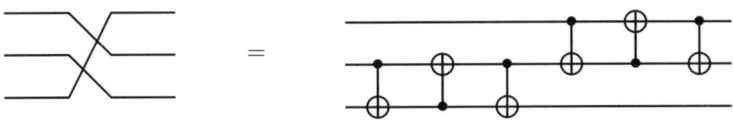

Fig. 4.2. Factorization $(1,3,2) = (1,2)(2,3)$

Example 4.2 (Controlled Operations). Following [172], we introduce a special class of quantum gates with multiple control qubits, yielding a natural generalization of the controlled NOT gate. This class of gates is given by the transformations $\Lambda_k(U)$, where U is a unitary transformation in $\mathcal{U}(2^l)$. The gate $\Lambda_k(U)$ is a transformation acting on $k+l$ qubits, where the k most significant bits serve as control bits and the l least significant bits are target bits: the operation U is applied to the l target bits if and only if all k control bits are equal to 1. Denoting by $M := 2^l(2^k-1)$ the number of basis vectors on which $\Lambda_k(U)$ acts trivially, the corresponding unitary matrix is given by $\mathbf{1}_M \oplus U$, where we have used $\oplus$ to denote a direct sum of matrices.

To provide further examples of the graphical notation for quantum circuits, we give in Fig. 4.3 a $\Lambda_1(U)$ gate for $U \in \mathcal{U}(2^n)$ with a normal control qubit, a gate $\overline{\Lambda}_1(U)$ with an inverted control qubit, and the matrices represented. Lemmas 7.2 and 7.5 of [172] show that for $U \in \mathcal{U}(2)$, the gate $\Lambda_k(U)$ can be realized with gate complexity $O(n)$, for $k < n-1$. If there are auxiliary qubits (so-called ancillae) available, a gate $\Lambda_{n-1}(U)$ can also be computed using $O(n)$ operations from $\mathcal{G}_1$; otherwise, we have $\kappa[\Lambda_{n-1}(U)] = O(n^2)$.

We remark that, if $U \in \mathcal{U}(2^n)$ can be realized in p elementary operations then, $\Lambda_1(U) \in \mathcal{U}(2^{n+1})$ can be realized in $c \times p$ basic operations, where $c \in \mathbb{N}$ is a constant that does not depend on U. To see this, we first assume, without loss of generality, that U is decomposed into elementary gates. Therefore,

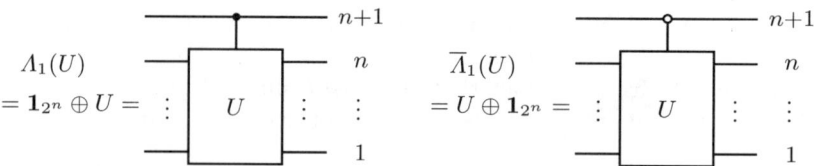

Fig. 4.3. Controlled gates with (*left*) normal and (*right*) inverted control bit. Here $\oplus$ is used to denote the block-direct sum of matrices

we have to show that a doubly controlled NOT (also called a Toffoli gate, see Sect. 4.2.3) and a singly controlled $U \in \mathcal{U}(2)$ gate can be realized with a constant increase of length. It is possible to obtain the bound $c \leq 17$ according to the following decompositions [172]: for each unitary transformation $U \in \mathcal{U}(2)$ we can write $\Lambda_1(U) = A^{(1)} \cdot \text{CNOT}^{(1,2)} \cdot B^{(1)} \cdot \text{CNOT}^{(1,2)} \cdot C^{(1)}$ with suitably chosen $A, B, C \in \mathcal{U}(2)$, i.e. we need at most five elementary gates for the realization of $\Lambda_1(U)$. To bound the number of operations necessary to realize $\tau := \Lambda_2(\sigma_x)$ with respect to the set $\mathcal{G}_1$, we choose a square root R of σ_x, i.e. $R^2 = \sigma_x$, and use the identity

$$\tau = [\mathbf{1}_2 \otimes \Lambda_1(R)] \cdot \text{CNOT}^{(2,3)} \cdot [\mathbf{1}_2 \otimes \Lambda_1(R^\dagger)] \cdot \text{CNOT}^{(2,3)} \cdot \Lambda_1(\mathbf{1}_2 \otimes R).$$

This shows that we need at most $5+1+5+1+5 = 17$ elementary gates to realize τ.

Example 4.3 (Cyclic Shift). Let $P_n \in S_{2^n}$ be the cyclic shift acting on the states of the quantum register as $x \mapsto x+1 \mod 2^n$. The corresponding permutation matrix is the 2^n-cycle $(0, 1, \ldots, 2^n-1)$. The unitary matrix P_n can be realized in a polylogarithmic number of operations; see Fig. 4.4 for a realization using Boolean gates only.

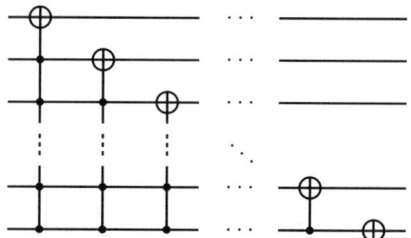

Fig. 4.4. Realizing a cyclic shift on a quantum register

Other, non-Boolean factorizations are also possible: using a basic fact about group circulants (see also Sect. 4.5.1) and anticipating the fact that the discrete Fourier transform DFT_{2^n} can be performed in $O(n^2)$ operations (which is shown in Sect. 4.4.1), we can use the identity

$$\mathrm{DFT}_{2^n}^{-1} \cdot P_n \cdot \mathrm{DFT}_{2^n} = \mathrm{diag}(\omega_{2^n}^i : i = 0, \ldots, 2^n - 1)$$
$$= \mathrm{diag}(1, \omega_{2^n}^{2^{n-1}}) \otimes \cdots \otimes \mathrm{diag}(1, \omega_{2^n})$$

to obtain $\kappa(P_n) = O(n^2)$.

4.2.2 Boolean Functions and the Ring Normal Form

Boolean functions are important primitives used throughout classical informatics. Denoting the finite field with q elements[2] by $GF(q)$, we obtain the Boolean numbers as the special case $q = 2$. A multivariate Boolean function $f : GF(2)^n \to GF(2)$ can be represented in various ways. Besides the truth table, which is a common but uneconomic way to represent f as the sequence of its values $f(0\ldots 0), \ldots, f(1\ldots 1)$ for all binary strings of length n, prominent examples of normal forms are the conjunctive and disjunctive normal forms [183] which originate from predicate logic and are used in transistor circuitry.

For quantum computational purposes, another way of representing f offers itself, namely the ring normal form (RNF), defined as the (unique) expansion of f as a polynomial in the ring R_n of Boolean functions of n variables. This ring is defined by $R_n := GF(2)[X_1, \ldots, X_n]/(X_1^2 - X_1, \ldots, X_n^2 - X_n)$ [184]. Multiplication and addition in R_n are the usual multiplication and addition of polynomials modulo the relations given by the ideal $(X_1^2 - X_1, \ldots, X_n^2 - X_n)$, and addition is usually denoted as "$\oplus$". Therefore f is represented as

$$f(X_1, \ldots, X_n) := \bigoplus_{u=(u_1,\ldots,u_n) \in \{0,1\}^n} a_u \prod_{i=1}^n X_i^{u_i}, \tag{4.1}$$

with coefficients $a_u \in GF(2)$.

Example 4.4. The logical complement is given by $\mathrm{NOT}(X) = 1 \oplus X$. The RNF of the AND function on n variables $X_1, \ldots, X_n$ is $\mathrm{AND}(X_1, \ldots, X_n) = \prod_{i=1}^n X_i$. The RNF of the PARITY function of n variables is given by $\mathrm{PARITY}(X_1, \ldots, X_n) = \bigoplus_{i=1}^n X_i$. Finally, the RNF of the OR function on n variables is given by

$$\mathrm{OR}(X_1, \ldots, X_n) = 1 \oplus \prod_{i=1}^n (1 \oplus X_i) = \bigoplus_m m(X_1, \ldots, X_n),$$

where the last sum runs over all nonconstant multilinear monomials m in the ring R_n.

---
[2] Necessarily, we have $q = p^n$, where p is a prime and $n \geq 1$ [182]. Finite fields are also called *Galois fields* after Évariste Galois (1811–1832).

We can implement a Boolean function given in the RNF shown in (4.1) with the use of the so-called Toffoli gate τ. The action of τ on the basis states of the Hilbert space $\mathcal{H}_8$ is given by $\tau : |x\rangle|y\rangle|z\rangle \mapsto |x\rangle|y\rangle|z \oplus x \cdot y\rangle$ [185].

Horner's rule for multivariate polynomials yields a method for implementing the function f given in (4.1). To achieve this, we write $f(X_1, \ldots, X_n) = f_1(X_1, \ldots, X_{n-1}) \cdot X_n \oplus f_2(X_1, \ldots, X_{n-1})$ and observe that this function can be computed using one Toffoli gate, assuming that f_1 and f_2 have already been computed. Therefore, we obtain a recursive factorization for f, which in general will make use of auxiliary qubits.

4.2.3 Embedded Transforms

This section deals with the issue of embedding a given transform A into a unitary matrix of larger size. We start by considering the problem of realizing a given matrix A as a submatrix of a unitary matrix of larger size. The following theorem (see also [186]) shows that the only condition A has to fulfill in order to allow an embedding involving one additional qubit is to be of bounded norm, i.e. $\|A\| \leq 1$, with respect to the spectral norm.

Theorem 4.2. *Let* $A \in \mathcal{C}^{n \times n}$ *be a given matrix of norm* $\|A\| \leq 1$. *Then*

$$U_A := \begin{pmatrix} A & (\mathbf{1}_n - AA^\dagger)^{1/2} \\ (\mathbf{1}_n - A^\dagger A)^{1/2} & -A^\dagger \end{pmatrix} \tag{4.2}$$

yields a unitary matrix $U_A \in \mathcal{U}(2n)$ *which contains* A *as the* $n \times n$ *submatrix in the upper left corner.*

Proof. Observe that the $n \times 2n$ matrix $U_1 := (A, (\mathbf{1}_n - A^\dagger A)^{1/2})^t$ has the property $U_1^\dagger \cdot U_1 = \mathbf{1}_n$. Analogously, for the matrix $U_2 := [A, (\mathbf{1}_n - AA^\dagger)^{1/2}]$, the identity $U_2 \cdot U_2^\dagger = \mathbf{1}_n$ holds. An easy computation shows that (4.2) is indeed unitary. □

Since each matrix in $\mathcal{C}^{n \times n}$ can be renormalized by multiplication with a suitable scalar to fulfill the requirement of a bounded norm, we can realize all operations up to a scalar prefactor by unitary embeddings. The embedding (4.2) is by no means unique. However, it is possible to parametrize *all* embeddings by $(\mathbf{1}_n \oplus V_1) \cdot U_A \cdot (\mathbf{1}_n \oplus V_2)$, where $V_1, V_2 \in \mathcal{U}(n)$ are arbitrary unitary transforms.

We are naturally led to a different kind of embedding if the given transformation is unitary and we want to realize it on a qubit architecture, i.e. if we restrict ourselves to matrices whose size is a power of 2. Then, a given unitary matrix $U \in \mathcal{U}(N)$ can be embedded into a unitary matrix in $\mathcal{U}(2^n)$ by choosing $n = \lceil \log N \rceil$ and padding U with an identity matrix $\mathbf{1}_{2^n - N}$ of size $2^n - N$. This is noncanonical since we have degrees of freedom in the choice of the subspace on which this newly formed matrix acts as the identity.

4 Quantum Algorithms: Applicable Algebra and Quantum Physics

In general, it is a difficult problem to find the optimal embedding for a given transform. An example in which it is not natural to go to the next power of 2 is given by $U = U_1 \otimes U_2 \in \mathcal{U}(15)$, where $U_1 \in \mathcal{U}(3)$ and $U_2 \in \mathcal{U}(5)$. We then have the possibilities $U \oplus 1_1 \in \mathcal{U}(2^4)$ and the embedding $(U_1 \oplus 1_1) \otimes (U_2 \oplus 1_3) \in \mathcal{U}(2^5)$, which respects the tensor decomposition of U.

A third type of embedding occurs in the context of quantum and reversible computing, where a general method is required to make a given map $f : X \to Y$ bijective (here X and Y are finite sets). If we consider the map $\tilde{f} : X \times Y \to X \times Y$ which maps $(x, y_0) \mapsto (x, f(x))$, where y_0 is a fixed element in the codomain Y of f, then this map is obviously injective when restricted to the fibre $X \times \{y_0\}$. Observe now that it is always possible to extend $\tilde{f}|_{X \times \{y_0\}}$ to a unitary operation on the Hilbert space $\mathcal{H}_{X,Y}$ spanned by the basis consisting of $\{|x\rangle|y\rangle : x \in X, y \in Y\}$. Hence, it is always possible to construct a unitary operation $V_f : \mathcal{H}_{X,Y} \to \mathcal{H}_{X,Y}$ which has the property

$$|x\rangle|0\rangle \mapsto |x\rangle|f(x)\rangle , \quad \text{for all } x \in X , \tag{4.3}$$

i.e. V_f implements the graph $\Gamma_f = \{(x, f(x)) : x \in X\}$ of f in the Hilbert space $\mathcal{H}_{X,Y}$. Here, we have identified the special element y_0 with the basis vector $|0\rangle \in \mathcal{H}_Y$.

Example 4.5. Let $f : GF(2)^2 \to GF(2)$ be the AND function, i.e. let $f = x \cdot y$ be the RNF of f. Note that $\tilde{f}$ can be chosen to be the function $(x, y, z) \mapsto (x, y, z \oplus f(x, y))$ since the codomain is endowed with a group structure. Overall, we obtain the function table of $\tilde{f}$ given in Fig. 4.5. The variables with a prime correspond to the values after the transformation has been performed.

x	y	z	x'	y'	z'
0	0	0	0	0	0
0	0	1	0	0	1
0	1	0	0	1	0
0	1	1	0	1	1

x	y	z	x'	y'	z'
1	0	0	1	0	0
1	0	1	1	0	1
1	1	0	1	1	1
1	1	1	1	1	0

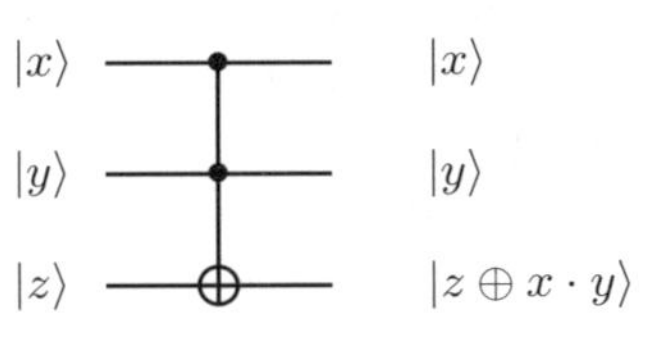

$|x\rangle \longrightarrow |x\rangle$
$|y\rangle \longrightarrow |y\rangle$
$|z\rangle \longrightarrow |z \oplus x \cdot y\rangle$

Fig. 4.5. Truth table for the Toffoli gate and the corresponding quantum circuit

We recognize $\tilde{f}$ as the unitary operation $\tau : |x\rangle|y\rangle|z\rangle \mapsto |x\rangle|y\rangle|z \oplus x \cdot y\rangle$ on the Hilbert space $\mathcal{H}_8$, which is the Toffoli gate [185].

The method described in Example 4.5 is quite general, as the following theorem shows (for a proof see [187]).

Theorem 4.3. *Suppose $f : \{0,1\}^n \to \{0,1\}^m$ is a Boolean function which can be computed using c operations from the universal set $\{\text{AND}, \text{NOT}\}$ of classical gates. Then $\tilde{f} : \{0,1\}^{n+m} \to \{0,1\}^{n+m}$, defined by $(x, y) \mapsto$*

$(x, y \oplus f(x))$, is a reversible Boolean function which can be computed by a circuit of length $2c + m$ built up from the set $\{\text{CNOT}, \tau\}$ of reversible gates.

Even though the construction described in Theorem 4.3 works for arbitrary $f : \{0,1\}^n \to \{0,1\}^m$, in general only $r_f = \lceil \log \max_{y \in \{0,1\}^m} |f^{-1}(y)| \rceil$ additional bits are necessary to define a reversible Boolean function $f_{\text{rev}} : \{0,1\}^{n+r_f} \to \{0,1\}^{n+r_f}$ with the property $f_{\text{rev}}|_{\{0,1\}^n} = f$. The reason is that by using the additional r_f bits, the preimages of f can be separated via a suitable binary encoding. However, the complexity of a Boolean circuit of a realization of f_{rev} constructed in such a way is such that the circuit cannot be controlled as easily as for the function defined in Theorem 4.3.

4.2.4 Permutations

We have already mentioned in Sect. 4.2.1 that on a quantum computer permutations of the basis states have to be taken into account when considering the complexity: in general, for the cost $\kappa(\pi)$, where π is a permutation matrix in $\mathcal{U}(2^n)$ and κ is the complexity measure introduced in Sect. 4.2.1, nothing better is known than an exponential upper bound of $O(n4^n)$.

Nevertheless, there are quite a few classes of permutations admitting a better, even polylogarithmic word length, as the examples of permutations of quantum wires and of the cyclic shift $P_n : x \mapsto x + 1 \bmod 2^n$ on a quantum register have shown (see Examples 4.1 and 4.3).

In what follows we consider a further class of permutations of a quantum register that admits efficient realizations, which operate by linear transformations on the *names* of the kets. Recall that the basis states can be identified with the binary words of length n and hence, with the elements of $GF(2)^n$, the n-dimensional vector space over the finite field $GF(2)$ of two elements. Denoting the group of invertible linear transformations of $GF(2)^n$ by $\text{GL}(n, GF(2))$, we see that each transform $A \in \text{GL}(n, GF(2))$ corresponds to a permutation of the binary words of length n and, hence, to a permutation matrix Φ_A of size $2^n \times 2^n$.

It turns out that these permutations are efficiently realizable on a quantum computer (see Sect. 4 of [188]). First we need the following lemma.

Lemma 4.1. *Let K be a field and let $A \in \text{GL}(n, K)$ be an invertible matrix with entries in K. Then there exist a permutation matrix P, a lower triangular matrix L and an upper triangular matrix U such that $A = P \cdot L \cdot U$.*

In numerical mathematics this decomposition is also known as the "LU decomposition" (see, e.g., Sect. 3.2. of [189]). The statement is a consequence of Gauss's algorithm. We are now ready to prove the following theorem.

Theorem 4.4. *Given $A \in \text{GL}(n, GF(2))$, Φ_A can be realized in $O(n^2)$ elementary operations on a quantum computer.*

Proof. First, decompose A according to Lemma 4.1 into $A = P \cdot L \cdot U$ and observe that the permutation matrix P is a permutation of the quantum wires, and hence $\kappa(P) = O(n)$ (see Example 4.1). The matrices L and U can be realized using CNOT gates only. Without loss of generality, we consider the factorization of L: proceeding along the diagonals of these matrices, we find all diagonal entries to be 1 (otherwise the matrices would not be invertible). Therefore Φ_L maps the basis vector $|e_i\rangle$, where $e_i = (0 \ldots 1 \ldots 0)$ is the ith basis vector in the standard basis of $GF(2)^n$, to the sum $\sum_{j \geq i} \alpha_j e_j$, where the vector $(\alpha_i)_{i=1,\ldots,n}$ is the ith column of A. Application of the sequence $\prod_{j \geq i} \text{CNOT}^{(i,j)}$, where the product runs over all $j \neq 0$, has the same effect on the basis vector e_i.

Proceeding column by column in L yields a factorization into $O(n^2)$ elementary gates. Combining the factorizations for P, L and U, we obtain $\kappa(A) = O(n^2)$. □

As an example, we take a look at the matrix

$$A = \begin{pmatrix} 1 & 0 \\ 1 & 1 \end{pmatrix} \in \text{GL}(2, GF(2)).$$

To see what the corresponding Φ_A looks like, we compute the effect of A on the basis vectors:

$$\begin{pmatrix} 0 \\ 0 \end{pmatrix} \mapsto \begin{pmatrix} 0 \\ 0 \end{pmatrix}, \quad \begin{pmatrix} 0 \\ 1 \end{pmatrix} \mapsto \begin{pmatrix} 0 \\ 1 \end{pmatrix}, \quad \begin{pmatrix} 1 \\ 0 \end{pmatrix} \mapsto \begin{pmatrix} 1 \\ 1 \end{pmatrix}, \quad \begin{pmatrix} 1 \\ 1 \end{pmatrix} \mapsto \begin{pmatrix} 1 \\ 0 \end{pmatrix},$$

i.e. $\Phi_A = \text{CNOT}^{(2,1)}$ in accordance with Theorem 4.4, since A is already lower triangular.

As further examples of permutations arising as unitary transforms on a quantum computer, we mention gates for modular arithmetic [190, 191]. More specifically, we consider the following operation, acting on kets which have been endowed with the group structure of $\mathbf{Z}_N := \mathbf{Z}/N\mathbf{Z}$, i.e. $\{|x\rangle : x \in \mathbf{Z}_N\}$ is a basis of this operational space $\mathcal{H}$. Then

$$\Upsilon_a : |x\rangle|0\rangle \mapsto |x\rangle|a \cdot x \bmod N\rangle,$$

where $a \in \mathbf{Z}_N^\times$ is an element of the multiplicative group of units in $\mathbf{Z}_N$; this can be extended to a permutation of the whole space $\mathcal{H} \otimes \mathcal{H}$ using the methods of Sect. 4.2.3. Using a number of ancilla qubits which is polynomial in $\log[\dim(\mathcal{H})]$, it is possible to realize Υ_a, as well as other basic primitives known from classical circuit design [183], such as

- adders modulo N: $|x\rangle|y\rangle \mapsto |x\rangle|x + y \bmod N\rangle$
- modular exponentiation: $|x\rangle|0\rangle \mapsto |x\rangle|a^x \bmod N\rangle$,

in polylogarithmic time on a quantum computer [190, 191].

4.2.5 Preparing Quantum States

If we are interested in preparing particular quantum states by means of an effective procedure, in most cases it is straightforward to write down a quantum circuit which yields the desired state when applied to the ground state $|0\rangle$. For instance, by means of the quantum circuit given in Fig. 4.6, a Schrödinger cat state $|\Psi_n\rangle$ on n qubits can be prepared using $n+1$ elementary gates. These are the states

$$|\Psi_n\rangle := \frac{1}{\sqrt{2}} \underbrace{|0\ldots 0\rangle}_{n \text{ zeros}} + \frac{1}{\sqrt{2}} \underbrace{|1\ldots 1\rangle}_{n \text{ ones}},$$

and we remind the reader that $|\Psi_2\rangle$ is locally equivalent to a so-called EPR state [24] and $|\Psi_3\rangle$ is a so-called GHZ state [37].

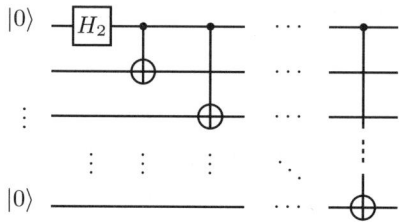

Fig. 4.6. Quantum circuit that prepares a cat state

We remark that there is an algorithm to prepare an arbitrary quantum state $|\varphi\rangle$ starting from the ground state $|0\rangle$, i.e. to construct a quantum circuit U_φ yielding $U_\varphi|0\rangle = |\varphi\rangle$.

Algorithm 1 *Let $|\varphi\rangle = \sum_{x \in \mathbf{Z}_2^n} \alpha_x |x\rangle$ be a quantum state which we would like to prepare. Do the following in a recursive way. Write*

$$|\varphi\rangle = a|0\rangle|\varphi_0\rangle + b|1\rangle|\varphi_1\rangle,$$

where a and b are complex numbers fulfilling $|a|^2 + |b|^2 = 1$. The states

$$|\varphi_0\rangle := \sum_{x \in \mathbf{Z}_2^{n-1}} \alpha_x^{(0)} |x\rangle \qquad \text{and} \qquad |\varphi_1\rangle := \sum_{x \in \mathbf{Z}_2^{n-1}} \alpha_x^{(1)} |x\rangle$$

appearing on the right-hand side of the above equation can be prepared by induction hypotheses using the circuits U_1 and U_2. Let A be the local transform

$$A := \begin{pmatrix} a & -\overline{b} \\ b & \overline{a} \end{pmatrix} \otimes \mathbf{1}_{2^{n-1}}.$$

Then $|\varphi\rangle$ can be prepared by application of the circuit $A \cdot \Lambda_1(U_1) \cdot \overline{\Lambda_1}(U_2)$ to the ground state $|0\rangle$.

4 Quantum Algorithms: Applicable Algebra and Quantum Physics   109

In general, the quantum circuit U_φ for preparing a state $|\varphi\rangle \in \mathcal{H}_{2^n}$ generated by this algorithm has a complexity $\kappa(U_\varphi) = O(2^n)$, which is linear in the dimension of the Hilbert space but exponential in the number of qubits. However, as the example of cat states previously mentioned shows, there are states which admit much more efficient preparation sequences. In such a set of states, we also find the so-called symmetric states [192]

$$\frac{1}{\sqrt{n+1}}(|00\ldots 0\rangle + |10\ldots 0\rangle + |01\ldots 0\rangle + \ldots + |00\ldots 1\rangle),$$

i.e. the union of the orbits of $|00\ldots 0\rangle$ and $|10\ldots 0\rangle$ under the cyclic group acting on the qubits. As shown in Sect. 4 of [192], these states can be prepared using $O(n)$ operations and a quadratic overhead of ancilla qubits.

Finally, we give circuits for preparation of the states $|\psi_\nu\rangle := (1/\sqrt{\nu})\sum_{i=1}^{\nu} |i\rangle$ for $\nu = 1, \ldots, 2^n$, which represent equal amplitudes over the first ν basis states of $\mathcal{H}_{2^n}$. The states $|\psi_\nu\rangle$ can be efficiently prepared from the ground state $|0\rangle$ by the following procedure (using the principle of binary search [193]), which is described in Sect. 4 of [187].

Since $|\psi_{2^n}\rangle$ can easily be prepared by application of the Hadamard transformation $H_2^{\otimes n}$, we can assume $\nu < 2^n$ without loss of generality. We now choose $k \in \mathbb{N}$ such that $2^k \leq \nu < 2^{k+1}$ and apply the transformation

$$U := \frac{1}{\sqrt{\nu}}\begin{pmatrix} \sqrt{2^k} & -\sqrt{\nu - 2^k} \\ \sqrt{\nu - 2^k} & \sqrt{2^k} \end{pmatrix}$$

to the first bit of the ground state $|0\rangle$. Next we achieve equal superposition on the first 2^k basis states $|0\ldots 0\rangle, \ldots, |0\ldots 01\ldots 1\rangle$ by application of an $(n-k)$-fold controlled $\overline{\Lambda}_1(H_2^{\otimes k})$ operation, which can be implemented using $O(n^2)$ operations. Finally, we apply the preparation circuit for the state $|\psi_{\nu-2^k}\rangle$ (which has been constructed by induction), conditioned on the $(k+1)$th bit. Overall, we obtain a complexity for the preparation of $|\psi_\nu\rangle$ of $O(n^3)$ operations.

4.2.6 Quantum Turing Machines

Quantum circuits provide a natural framework to specify unitary transformations on finite-dimensional Hilbert spaces and give rise to complexity models when factorizations into elementary gates, e.g. with respect to the universal sets $\mathcal{G}_1$ or $\mathcal{G}_2$, are taken into account.

Besides the formalism of quantum networks, there are other ways of describing computations performed by quantum mechanical systems. In the following we briefly review the model of quantum Turing machines (QTMs) defined by Deutsch [7]. We remind the reader that Turing machines [194] provide a unified model for *classical* deterministic and probabilistic computation (see, e.g., [195]). The importance of Turing machines as a unifying concept

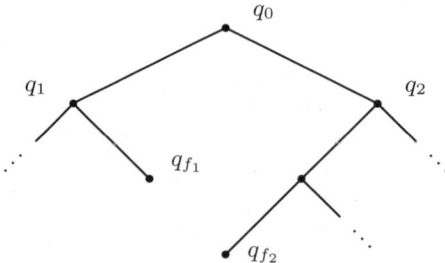

Fig. 4.7. Configurations of a Turing machine

for classical computing manifests itself in the Church–Turing thesis [196, 197], which, in its modern form, claims that every physically reasonable model of computation can be efficiently simulated on a probabilistic Turing machine.

Definition 4.2. *A deterministic Turing machine T is defined by the data (Q, Σ, q_0, F, t), where Q is a finite set of states, Σ a set of symbols, $q_0 \in Q$ a distinguished initial state, $F \subseteq Q$ the set of final states, and $t : Q \times \Sigma \longrightarrow Q \times \Sigma \times \{\leftarrow, \downarrow, \rightarrow\}$ the transition function. The admissible actions of T are movements of a read–write head, which in one computational step can move to the left, stay where it is or move to the right (we have denoted these actions by $\{\leftarrow, \downarrow, \rightarrow\}$). Along with the Turing machine T comes a infinite tape of cells (the cells are in bijection with $\mathbf{Z}$) which can take symbols from Σ.*

A *configuration* of a Turing machine T is therefore given by a triplet $(v, p, q) \in (\Sigma^{\mathbf{Z}}, \mathbf{Z}, Q)$, consisting of the state v of the tape, the position p of the head and the internal state q. We obtain a tree of configurations by considering two configurations c_1 and c_2 to be adjacent if and only if c_2 is obtained from c_1 by an elementary move, i.e. scanning a symbol from the tape, changing the internal state, writing back a symbol to the tape and moving the head. The initial state is the root of this tree, whereas the final states constitute its leaves (see Fig. 4.7).

A *probabilistic* Turing machine differs from a deterministic one only in the nature of the transition function t, which is then a mapping

$$t : Q \times \Sigma \times Q \times \Sigma \times \{\leftarrow, \downarrow, \rightarrow\} \longrightarrow [0, 1],$$

which assigns probabilities from the real interval $[0, 1]$ to the possible actions of T. A normalization condition, which guarantees the well-formedness of a probabilistic Turing machine, is that for all configurations the sum of the probabilities of all successors is 1. Therefore, the admissible state transitions of a probabilistic Turing machine T can be described by a stochastic matrix $S_T \in [0, 1]^{\mathbf{Z} \times \mathbf{Z}}$, where stochasticity means that the rows of S_T add up to 1 and the successor c_{succ} of c is obtained by $c_{\text{succ}} = S_T \cdot c$. Note that a deterministic Turing machine is a special case of a probabilistic Turing machine with a "subpermutation" matrix S_T, i.e. S_T can be obtained from a suitable permutation matrix by deleting some of its rows.

One more variation of this idea is needed to finally arrive at the concept of a *quantum* Turing machine: we have to require that the transition function t is a mapping with a normalization condition

$$t : Q \times \Sigma \times Q \times \Sigma \times \{\leftarrow, \downarrow, \rightarrow\} \longrightarrow \mathcal{C} \tag{4.4}$$

from a configuration to possibly many successors, each of which is given a complex amplitude. Here the normalization constraint says that the matrix U_T describing the dynamics of T on the state space is *unitary*.

Observe that there is one counterintuitive fact implied by this definition: as t assigns complex amplitudes to $Q \times \Sigma \times Q \times \Sigma \times \{\leftarrow, \downarrow, \rightarrow\}$ (according to (4.4)), one can interpret a configuration of T as being in a superposition of (i) tape symbols in each individual cell, (ii) states of the finite-state machine supported by Q and (iii) positions of the head. The last point in particular might look uncomfortable at first sight, but we remind the reader of the fact that in classical probabilistic computation each individual configuration is assigned a probability, so one can think of a probabilistic computation as traversing an exponentially large configuration space! The main difference of the QTM model is that because of negative amplitudes, computational paths in this configuration space can cancel each other out, i.e. the effects of interference can force the Turing machine into certain paths which ultimately may lead to the desired solution of the computational task.

Now that the computational model of a QTM has been defined, the question arises as to what can be computed on a QTM, compared with a classical deterministic or probabilistic Turing machine. An important result in this context is that everything which can be computed classically in polynomial time can also be computed on a QTM because of the following theorem, which relies on some results of Bennett for reversible Turing machines [55, 198, 199] and was adapted to the QTM setting in [200]. As usual, we denote by L^* the language $\{0, 1\}^*$ consisting of all binary strings and denote by $|x|$ the length of the word $x \in L^*$.

Theorem 4.5. *Let $f : L^* \to L^*$ be a polynomial-time computable function such that $|f(x)|$ depends only on $|x|$. Then there is a polynomial-time QTM T that computes $|x\rangle|0\rangle \mapsto |x\rangle|f(x)\rangle$. The running time of T depends only on $|x|$.*

Proof. The basic idea is to replace each elementary step in the computation of f by a reversible operation (using theorem 4.3), keeping in mind that ancilla qubits are needed to make the computation reversible (see Sect. 4.2.3). We now adjoin additional qubits to the system, which are initialized in the ground state $|0\rangle$, and apply a controlled NOT operation using the computational qubits holding the result $f(x)$. Of course, after the application of this operation the state is highly entangled between the computational register and the additional register holding the result. Next, we run the whole computation that was done to compute f, backwards, reversibly, on the computation register to get rid of the garbage which might destroy the coherence, and end up with the state $|x, 0, f(x)\rangle$, where the $|0\rangle$ refers to the ancilla bits used in the first step of this procedure. □

Remark 4.2.

- The class of quantum Turing machines allows the definition and study of the important complexity class BQP [200],[3] as well as the relation of BQP to other classes known from classical complexity theory.
- There are programming primitives for QTMs, such as composition, loops and branching [200], as in the classical case. However, a problem arises in realizing while-loops, since the predicate which decides whether the loop terminates can be in a superposition of true and false, depending on the computation path. Therefore all computations have to be arranged in such a way that this predicate is never in a superposed state, i.e. the state of the predicate has to be classical. As a consequence, we obtain the result that a quantum Turing machine can only perform loops with a prescribed number of iterations, which in turn can be determined by a classical Turing machine.
- An important issue is whether QTMs constitute an analog or discrete model of computation. One might be tempted to think of the possibility of encoding an arbitrary amount of information into the transition amplitudes of t, i.e. of producing a machine model which could benefit from computing with complex numbers to arbitrary precision (for the strange effects of such models see, e.g., [201]). However, see [200, 202] for a proof of the fact that it is sufficient to take transition amplitudes from the finite set $\{\pm 3/5, \pm 4/5, \pm 1, 0\}$ in order to approximate a given QTM to arbitrary precision. The reason for this is that the Pythagorean-triple transformation

$$\frac{1}{5}\begin{pmatrix} 3 & -4 \\ 4 & 3 \end{pmatrix} \in \mathrm{SO}(2)$$

has eigenvalues of the form $e^{2\pi i \nu}$ with $\nu \notin \mathbb{Q}$ and, therefore, generates a dense subgroup in $\mathrm{SO}(2)$. The statement then follows from the fact that the full unitary group on $\mathcal{H}_{2^n}$ can be parametrized by $\mathrm{SO}(2)$ matrices applied to arbitrary basis states and phase rotations [172, 203].
- Yao has shown [204] that the computational models of QTM and uniform families of quantum gates (see Sect. 4.2.1) are polynomially equivalent, i.e. each model can simulate the other with polynomial time overhead.

---
[3] BQP stands for "bounded-error quantum polynomial time".

4.3 Using Entanglement for Computation: A First Quantum Algorithm

Entanglement between registers holding quantum states lies at the heart of the quantum algorithm which we shall describe in this section. The formulation of a problem on which a quantum computer will exceed the performance of any classical (probabilistic) computer may appear artificial. However, this was one of the first examples of problems on which a quantum computer could provably outperform any classical computer, with an exponential speedup.

Because of its clarity and methodology, we present the quantum algorithm of Simon, in which many of the basic principles of quantum computing, namely the superposition principle, computing with preimages and the use of the Fourier transform, become apparent. We briefly remind the reader of the problem and mention that we are considering here a slightly generalized situation compared with the original setup (see also [57, 205, 206]).

Quantum algorithms relying on the same principles have been given in [56, 200]. As in the case of Simon's problem described below, these algorithms rely on the Fourier transform for a suitably chosen abelian group. In both cases it has been shown that these quantum algorithms provide a superpolynomial gap over any classical probabilistic computer in the number of operations necessary to solve the corresponding problems.

In the following, we denote by $\mathbf{Z}_2^n$ the elementary abelian 2-group of order 2^n, the elements of which we think of as being identified with binary strings of length n, and denote addition in $\mathbf{Z}_2^n$ by $\oplus$.

Definition 4.3 (Simon's Problem). *Let $f : \mathbf{Z}_2^n \to \mathbf{Z}_2^n$ be a function given as a black-box quantum circuit, i.e. f can be evaluated on superpositions of states and is realized by a unitary transform V_f specified by*

$$|x\rangle|0\rangle \mapsto |x\rangle|f(x)\rangle, \quad \text{for all } x \in \mathbf{Z}_2^n ,$$

as described in Sect. 4.2.3 (see, in particular, (4.3)). In addition, it is specified that there is a subgroup $U \subseteq \mathbf{Z}_2^n$ (the "hidden" subgroup) such that f takes a constant value on each of the cosets $g \oplus U$ for $g \in \mathbf{Z}_2^n$ and, furthermore, f takes different values on different cosets. The problem is to find generators for U.

We can now formulate a quantum algorithm which solves Simon's problem in a polynomial number of operations on a quantum computer. This algorithm uses $O(n)$ evaluations of the black-box quantum circuit f, and the classical postcomputation, which is essentially linear algebra over $GF(2)$, and also takes a number of operations which is polynomial in n.

Algorithm 2 *This algorithm needs two quantum registers of length n, holding elements of the domain and codomain of f, and consists of the following steps.*

1. Prepare the ground state

$$|\varphi_1\rangle = |0\ldots 0\rangle \otimes |0\ldots 0\rangle$$

in both quantum registers.

2. Achieve equal amplitude distribution in the first register, for instance by an application of a Hadamard transformation to each qubit:

$$|\varphi_2\rangle = \frac{1}{\sqrt{2^n}} \sum_{x \in \mathbf{Z}_2^n} |x\rangle \otimes |0\ldots 0\rangle \ .$$

3. Apply V_f to compute f in superposition. We obtain

$$|\varphi_3\rangle = \frac{1}{\sqrt{2^n}} \sum_{x \in \mathbf{Z}_2^n} |x\rangle |f(x)\rangle \ .$$

4. Measure the second register to obtain some value z in the range of f. Owing to the condition on f specified, the first register now holds a coset $g_0 \oplus U$ of the hidden subgroup U, namely the set of elements equal to $z = f(g_0)$:

$$|\varphi_4\rangle = \frac{1}{\sqrt{|U|}} \sum_{f(x)=z} |x\rangle|z\rangle = \frac{1}{\sqrt{|U|}} \sum_{x \in g_0 \oplus U} |x\rangle|z\rangle \ .$$

5. Application of the Hadamard transformation $H_2^{\otimes n}$ to the first register transforms the coset into the superposition $\sum_{y \in U^\perp} (-1)^{y \cdot g_0} |y\rangle$. The supported vectors of this superposition are the elements of $U^\perp$, which is the group defined by $U^\perp := \{y \in \mathbf{Z}_2^n : x \cdot y = \sum_{i=1}^n x_i y_i = 0\}$, i.e. the orthogonal complement of U with respect to the scalar product in $\mathbf{Z}_2^n$ (see also Sect. 4.4.2).

6. Now measure the first register. We draw from the set of irreducible representations of $\mathbf{Z}_2^n$ having U in the kernel, i.e. we obtain an equal distribution over the elements of $U^\perp$.

7. By iterating steps 1-6, we produce elements of $\mathbf{Z}_2^n$ which generate the group $U^\perp$ with high probability. After performing this experiment an expected number of n times, we generate with probability greater than $1-2^{-n}$ the group $U^\perp$.

8. By solving linear equations over $GF(2)$, it is easy to find generators for

$$(U^\perp)^\perp = U \ .$$

Therefore we obtain generators for U by computing the kernel of a matrix over $GF(2)$ in time $O(n^3)$.

Analysis of Algorithm 2

- We first address the measurement in step 4. If we do not perform this measurement, we are left with the state

$$|\varphi_4'\rangle = \frac{1}{\sqrt{2^n}} \sum_{\sigma \in \mathbf{Z}_2^n/U} \sum_{x \in U} |\sigma \oplus x\rangle |f(\sigma)\rangle \;,$$

and if we continue with step 5 we shall obtain the state

$$|\varphi_5'\rangle = \frac{1}{\sqrt{2^n}} \sum_{\sigma \in \mathbf{Z}_2^n/U} \sum_{y \in U^\perp} (-1)^{\sigma \cdot y} |y\rangle |f(\sigma)\rangle \;.$$

Therefore, sampling of the first register as in step 6 will yield an equal distribution over $U^\perp$ and we can go on as in step 7. Hence we can omit step 4.

- The reason for the application of the transformation $H_2^{\otimes n}$ and the appearance of the group $U^\perp$ will be clarified in the following sections. As it turns out, $H_2^{\otimes n}$ is an instance of a Fourier transform for an abelian group and $\perp$ is an antiisomorphism of the lattice of subgroups of $\mathbf{Z}_2^n$.

- For the linear-algebra part in step 8, we refer the reader to standard texts such as [207]. Gauss's algorithm for computing the kernel of an $n \times n$ matrix takes $O(n^3)$ arithmetic operations over the finite field $\mathbf{Z}_2 = GF(2)$. Overall, we obtain the following cost: $O(n)$ applications of V_f, $O(n^2)$ elementary quantum operations (which are all Hadamard operations H_2), $O(n^2)$ measurements of individual qubits, and $O(n^3)$ classical operations (arithmetic in $GF(2)$).

4.4 Quantum Fourier Transforms: the Abelian Case

In this section we recall the definition and basic properties of the discrete Fourier transform (DFT) and give examples of its use in quantum computing.

In most standard texts on signal processing (e.g. [208]), the DFT_N of a periodic signal given by a function $f : \mathbf{Z}_N \to \mathcal{C}$ (where $\mathbf{Z}_N$ is the cyclic group of order N) is defined as the function F given by

$$F(\nu) := \sum_{t \in \mathbf{Z}_N} e^{2\pi i \nu t/N} f(t) \;.$$

If, equivalently, we adopt the point of view that the signal f and the Fourier transform are vectors in $\mathcal{C}^N$, we see that performing the DFT_N is a matrix vector multiplication of f with the unitary matrix

$$\mathrm{DFT}_N := \frac{1}{\sqrt{N}} \cdot \left[\omega^{i \cdot j}\right]_{i,j=0,\ldots,N-1},$$

where $\omega = e^{2\pi i/N}$ denotes a primitive Nth root of unity.

From an algebraic point of view, the DFT_N gives an isomorphism Φ of the group algebra $\mathcal{C}\mathbf{Z}_N$,

$$\Phi : \mathcal{C}\mathbf{Z}_N \longrightarrow \mathcal{C}^N ,$$

onto the direct sum of the irreducible matrix representations of $\mathbf{Z}_N$ (where multiplication is performed pointwise). This means that DFT_N decomposes the regular representation of $\mathbf{Z}_N$ into its irreducible constituents. It is known that this property allows the derivation of a fast convolution algorithm [171] in a canonical way and can be generalized to more general group circulants. Viewing the DFT as a decomposition matrix for the regular representation of a group leads to the generalization of Fourier transforms to arbitrary finite groups (cf. [171] and Sect. 4.6.1).

The fact that DFT_N can be computed in $O(N \log N)$ arithmetic operations (counting additions and multiplications) is very important for applications in classical signal processing. This possibility to perform a fast Fourier transform justifies the heavy use of the DFT_N in today's computing technology. In the next section we shall show that the $O(N \log N)$ bound, which is sharp in the arithmetic complexity model (see Chap. 4 and 5 of [174]) can be improved with a quantum computer to $O[(\log N)^2]$ operations.

4.4.1 Factorization of DFT_N

From now on we restrict ourselves to cases of DFT_N where $N = 2^n$ is a power of 2, since these transforms naturally fit the tensor structure imposed by the qubits.

The efficient implementation of the Fourier transform on a quantum computer starts from the well-known Cooley–Tukey decomposition [209]: after the row permutation Π_τ, where $\tau = (1,\ldots,n)$ is the cyclic shift on the qubits, is performed, the DFT_{2^n} has the following block structure [171]:

$$\Pi_\tau \, \mathrm{DFT}_{2^n} = \left(\begin{array}{c|c} \mathrm{DFT}_{2^{n-1}} & \mathrm{DFT}_{2^{n-1}} \\ \hline \mathrm{DFT}_{2^{n-1}} W_n & -\mathrm{DFT}_{2^{n-1}} W_n \end{array} \right)$$

$$= (\mathbf{1}_2 \otimes \mathrm{DFT}_{2^{n-1}}) \cdot T_n \cdot (\mathrm{DFT}_2 \otimes \mathbf{1}_{2^{n-1}}) .$$

Here we denote by

$$T_n := \mathbf{1}_{2^{n-1}} \oplus W_n, \quad W_n := \begin{pmatrix} 1 & & & & \\ & \omega_{2^n} & & & \\ & & \omega_{2^n}^2 & & \\ & & & \ddots & \\ & & & & \omega_{2^n}^{2^{n-1}} \end{pmatrix}$$

4 Quantum Algorithms: Applicable Algebra and Quantum Physics

the matrix of *twiddle factors* [171]. Taking into account the fact that W_n has the tensor decomposition

$$W_n = \bigotimes_{i=0}^{n-2} \begin{pmatrix} 1 & \\ & \omega_{2^n}^{2^i} \end{pmatrix},$$

we see that T_n can be implemented by $n-1$ gates having one control wire each. These can be factored into the elementary gates $\mathcal{G}_1$ with constant overhead.

Because tensor products are free in our computational model, by recursion we arrive at an upper bound of $O(n^2)$ for the number of elementary operations necessary to compute the discrete Fourier transform on a quantum computer (this operation will be referred to as "QFT").

In Fig. 4.8, the derived decomposition into quantum gates is displayed using the graphical notation introduced in Sect. 4.2.1. The gates labeled by D_k in this circuit are the diagonal phase shifts $\mathrm{diag}(1, e^{2\pi i/k})$ and, in addition, we have used the abbreviation $N = 2^n$. We observe that the permutations Π_n, which arose in the Cooley–Tukey formula, have all been collected together, yielding the so-called bit reversal, which is the permutation of the quantum wires $(1, n)(2, n-1)\ldots(n/2, n/2 + 1)$ when n is even and $(1, n)(2, n-1)\ldots((n-1)/2, (n+3)/2)$ when n is odd.

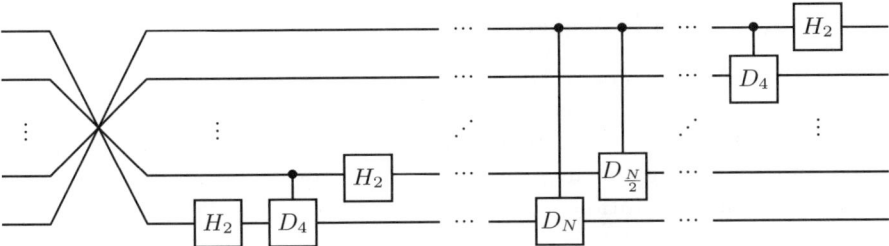

Fig. 4.8. Quantum circuit that computes a Fourier transform QFT_{2^n}

If we intend to use the Fourier transform as a sampling device, i.e. the application of QFT_{2^n} to a state $|\psi\rangle$ followed directly by a measurement in the standard basis, we can use the structure of this quantum circuit that computes a Fourier transform QFT_{2^n} to avoid nearly all quantum gates [210].

In order to give circuits for the QFT for an arbitrary abelian group, we need to describe the structure of these groups and their representations first, which is done in the next section.

4.4.2 Abelian Groups and Duality Theorems

Let A be a finite abelian group. Then A splits into a direct product of its p-components: $A \cong A_{p_1} \times \ldots \times A_{p_n}$, where $p_i, i = 1, \ldots, n$ are the prime divisors of the order $|A|$ of A (see [211], Part I, paragraph 8).

Furthermore, each A_{p_i}, which consists of all elements annihilated by a power of p_i, has the form $A_{p_i} \cong \mathbf{Z}_{p^{r_{i,1}}} \times \ldots \times \mathbf{Z}_{p^{r_{i,k_i}}}$ (see [211], Part I, paragraph 8). Both statements follow immediately from the structure theorem for finitely generated modules over principle ideal rings.

We remark that a Fourier transform for a finite abelian group can easily be constructed from this knowledge: given two groups G_1 and G_2, the irreducible representations of their direct product $G_1 \times G_2$ can be obtained from those of G_1 and G_2 as follows.

Theorem 4.6. *The irreducible representations of $G = G_1 \times G_2$ are given by*

$$\mathrm{Irr}(G) = \{\phi_1 \otimes \phi_2 : \phi_1 \in \mathrm{Irr}(G_1), \phi_2 \in \mathrm{Irr}(G_2)\} \ .$$

(For a proof see [212].) If we therefore encode the elements of A according to the direct-product decomposition as above, the matrix

$$\mathrm{DFT}_A = \bigotimes_{i=1}^{n} \bigotimes_{j=1}^{k_i} \mathrm{DFT}_{p_i^{r_{i,j}}}$$

is a decomposition matrix for the regular representation of A.

Corollary 4.1. *Let A be a finite abelian group of order 2^n. Then a Fourier transform for A can be computed in $O(n^2)$ elementary operations.*

Proof. The decomposition of DFT_{2^n} has already been considered and an implementation in $O(n^2)$ many operations has been derived from the Cooley–Tukey formula in Sect. 4.4.1. Since tensor products are free in our computational model, we can conclude that a direct factor $\mathbf{Z}_{2^n}$ is already the worst case for an implementation. □

Example 4.6. The Fourier transform for the elementary abelian 2-group $\mathbf{Z}_2^n$ is given by the tensor product $\mathrm{DFT}_2 \otimes \cdots \otimes \mathrm{DFT}_2$ of the Fourier transform for the cyclic factors and hence coincides with the Hadamard matrix $H_2^{\otimes n}$ used in Algorithm 2.

The Dual Group Given a finite abelian group A, we can consider $\mathrm{Hom}(A, \mathcal{C}^*)$, i.e. the group of characters of A (NB: in the nonabelian case a character is generalized to the traces of the representing matrices and hence is not a homomorphism anymore). The following theorem says that A is isomorphic to its group of characters.

Theorem 4.7. *For a finite abelian group A, we have $A \cong_\phi \mathrm{Hom}(A, \mathcal{C}^*)$.*

We shall make the isomorphism ϕ explicit. Choose $\omega_{p_1}, \ldots, \omega_{p_n}$, where the ω_i are primitive p_ith roots of unity in $\mathcal{C}^*$. Then ϕ is defined by the assignment $\phi(e_i) := \omega_{p_i}$ on the elements $e_i := (0, \ldots, 1, \ldots, 0)$, where the 1 is in the ith position. The isomorphism ϕ is not canonical, since a different choice of primitive roots of unity will yield a different isomorphism.

Next we observe that there is a pairing β (i.e. a bilinear map) between A and $\mathrm{Hom}(A, \mathcal{C}^*)$ via $\beta : (a, f) \mapsto f(a) \in \mathcal{C}^*$. We suppose that a is orthogonal to f (in symbols, $a \perp f$) if $\beta(a, f) = 1$. For a given subgroup $U \subseteq A$ we can now define (see also Algorithm 2) the orthogonal complement $U^\perp := \{y \in A : \beta(y, x) = 1, \quad \forall x \in U\}$. The following duality theorem holds (see [184]):

Theorem 4.8. *For all subgroups $U, U' \subseteq A$, the following identities hold:*

(a) Self-duality of $\perp$:
$$(U^\perp)^\perp = U .$$

(b) Complementarity:
$$(U \cap U')^\perp = \langle U^\perp, U'^\perp \rangle .$$

(c) The mapping $\perp$ is an inclusion-reversing antiisomorphism on the lattice of subgroups of A (i.e. $\perp$ is a Galois correspondence).

4.4.3 Sampling of Fourier Coefficients

In this section we address the problem of gaining information from the Fourier coefficients of a special class of states. More precisely, we consider the Fourier transforms of the (normalized) characteristic function

$$|\chi_{c+U}\rangle := \frac{1}{\sqrt{|U|}} \sum_{x \in c+U} |x\rangle \tag{4.5}$$

of a coset $c + U$ of a subgroup $U \subseteq A$ of an abelian group A. The question of what conclusions about U can be drawn from measuring the Fourier transformed state (4.5) is of special interest, as we have already seen in Sect. 4.3. The following theorem shows that for an abelian group, the Fourier transform maps subgroups to their duals.

Theorem 4.9. *Let $\mathrm{DFT}_A = (1/\sqrt{|A|}) \sum_{x,y \in A} \beta(x, y)|y\rangle \langle x|$ be the Fourier transform for the abelian group A. Then for each subgroup $U \subseteq A$, we have*

$$\mathrm{DFT}_A \frac{1}{\sqrt{|U|}} \sum_{x \in U} |x\rangle = \frac{1}{\sqrt{|U^\perp|}} \sum_{y \in U^\perp} |y\rangle .$$

Proof. Since

$$\mathrm{DFT}_A \frac{1}{\sqrt{|U|}} \sum_{x \in U} |x\rangle = \frac{1}{\sqrt{|U|}} \left(\sum_{x,y \in A} \beta(x, y)|y\rangle \langle x| \right) \sum_{x \in U} |x\rangle$$

$$= \frac{1}{\sqrt{|A||U|}} \sum_{y \in A} \sum_{x \in U} \beta(x, y)|y\rangle ,$$

it suffices to show that $\sum_{x \in U} \beta(x,y) = 0$ for $y \notin U^\perp$, but this statement follows from the fact that the existence of x_0 with $\beta(x_0, y) \neq 0$ implies that $\sum_{x \in U} \beta(x,y) = \sum_{x \in U} \beta(x+x_0, y) = \beta(x_0, y) \sum_{x \in U} \beta(x,y)$. The other case, $\sum_{x \in U} \beta(x,y) = |U|$ for $y \in U^\perp$, is obvious. □

Hence, measuring the Fourier spectrum of $|\chi_U\rangle$ yields an equal distribution on the elements of the dual group $U^\perp$. Also, in the case of the characteristic function of a coset $c + U$ instead of U, we obtain the same probability distribution on $U^\perp$ since the Fourier transform diagonalizes the group action completely in the abelian case, i.e. the translation by c corresponds to a pointwise multiplication by phases in the Fourier basis:

$$\mathrm{DFT}_A \frac{1}{\sqrt{|U|}} \sum_{x \in c+U} |x\rangle = \frac{1}{\sqrt{|U^\perp|}} \sum_{y \in U^\perp} \varphi_{c,y} |y\rangle ,$$

where $\varphi_{c,y} \in \mathcal{U}(1)$ are phase factors which depend on c and y but are always eth roots of unity, where e denotes the exponent of A. Since making measurements involves taking the squares of the amplitudes, we obtain an equal distribution over $U^\perp$. The states $|\chi_{c+U}\rangle$, which in general will be highly entangled, make the principle of *interference* and its use in quantum algorithms apparent: only those Fourier coefficients remain which correspond to the elements of $U^\perp$ (constructive interference), whereas the amplitudes of all other elements vanish (destructive interference).

4.4.4 Schur's Lemma and its Applications in Quantum Computing

In this section we explore the underlying reason behind Theorem 4.9, namely Schur's lemma. We present further applications of this powerful tool from representation theory. For a proof of Schur's lemma we refer the reader to Sect. 2 of [213].

Lemma 4.2 (Schur's lemma). *Let $\rho_1 : G \to \mathrm{GL}_\mathcal{C}(V)$ and $\rho_2 : G \to \mathrm{GL}_\mathcal{C}(W)$ be irreducible complex representations of a group G. Suppose that the element $A \in \mathrm{End}(V, W)$ has the property*

$$\rho_1(g) \cdot A = A \cdot \rho_2(g), \forall g \in G ,$$

i.e. A commutes with all pairs of images $\rho_1(g), \rho_2(g)$. Then exactly one of the following cases holds:

(i) $\rho_1 \not\cong \rho_2$. In this case $A = \mathbf{0}_{\mathrm{End}(V,W)}$.
(ii) $\rho_1 \cong \rho_2$. In this case A is a homothety, i.e. $A = \lambda \cdot \mathbf{1}_{\mathrm{End}(V,W)}$ with an element $\lambda \in \mathcal{C}$.

We give two applications of Schur's lemma which make its importance in the context of Fourier analysis apparent. First, we present a reformulation of Theorem 4.9 in representation-theoretical terms and state then a theorem which is useful in the sampling of functions having a hidden normal subgroup.

4 Quantum Algorithms: Applicable Algebra and Quantum Physics

Theorem 4.10. *Let G be a finite abelian group, $\operatorname{Hom}(G, \mathcal{C}^*)$ the dual group, and $\beta : G \times \operatorname{Hom}(G, \mathcal{C}^*) \to \mathcal{C}^*$ the canonical pairing defined by $\beta(g, \varphi) := \varphi(g)$. Then for each subgroup $U \subseteq G$, the following holds (normalization omitted):*

$$\operatorname{DFT}_G \sum_{u \in U} |u\rangle = \sum_{\substack{\varphi \in \operatorname{Hom}(G, \mathcal{C}^*) \\ U \subseteq \operatorname{Ker}(\varphi)}} |\varphi\rangle \ .$$

Moreover, we obtain the following identity for the cosets $u_0 + U$:

$$\operatorname{DFT}_G \sum_{u \in U} |u_0 + u\rangle = \sum_{\substack{\varphi \in \operatorname{Hom}(G, \mathcal{C}^*) \\ U \subseteq \operatorname{Ker}(\varphi)}} \beta(u_0, \varphi) \cdot |\varphi\rangle \ .$$

Proof. The mapping $\operatorname{DFT}_G : \mathcal{C}G \to \bigoplus_{i=1}^{|G|} \mathcal{C}$ is given by evaluation of elements of G for the irreducible representations $\{\varphi_1, \ldots, \varphi_s\}$ of G, which are all one-dimensional (i.e. $s = |G|$) and hence are characters, since G was assumed to be abelian. Therefore, the coefficient for the irreducible representation φ_i is computed from

$$\sum_{u \in U} \varphi_i(u_0 + u) = \sum_{u \in U} \varphi_i(u_0) \cdot \varphi_i(u)$$

$$= \varphi_i(u_0) \cdot \sum_{u \in U} \varphi_i(u) \ .$$

Considering the restricted characters $\varphi_i \downarrow U$, we use Schur's lemma (Lemma 4.2) to deduce that $\sum_{u \in U} \varphi_i(u) = 0$ iff $U \not\subseteq \operatorname{Ker}(\varphi_i)$. □

Theorem 4.11. *Let G be an arbitrary finite group and $N \triangleleft G$ a normal subgroup of G. Then for each irreducible representation ρ of G of degree d, exactly one of the following cases holds:*

$$\frac{1}{|N|} \sum_{n \in N} \rho(n) = \lambda \cdot \mathbf{1}_{d \times d} \ , \quad \text{or} \quad \frac{1}{|N|} \sum_{n \in N} \rho(n) = \mathbf{0}_{d \times d} \ ,$$

where the first case applies iff N is contained in the kernel of ρ.

Proof. Let $A := (1/|N|) \sum_{n \in N} \rho(n)$ denote the equal distribution over all images of N under ρ. Then, from the assumption of normality of N, we obtain

$$\rho(g)^{-1} \cdot A \cdot \rho(g) = A \ ,$$

i.e. A commutes with the irreducible representation ρ. Using Schur's lemma, we conclude that $A = \lambda \cdot \mathbf{1}_{d \times d}$. If $N \subseteq \operatorname{Ker}(\rho)$ we find $A = \mathbf{1}_{d \times d}$. On the other hand, if $N \not\subseteq \operatorname{Ker}(\rho)$, we conclude that A is the zero matrix $\mathbf{0}_{d \times d}$, because otherwise an element $n_0 \notin N$ equal to $\rho(n_0) \neq \mathbf{1}_{d \times d}$ would lead to the contradiction

$$\rho(n_0) \cdot \sum_{n \in N} \rho(n) = \sum_{n \in N} \rho(n_0 \cdot n) = \sum_{n \in N} \rho(n),$$

since from this we could conclude $\rho(n_0) = \mathbf{1}_{d \times d}$, contrary to the assumption $n_0 \notin \text{Ker}(\rho)$. □

4.5 Exploring Quantum Algorithms

4.5.1 Grover's Algorithm

We give an outline of Grover's algorithm for searching an unordered list and present an optical implementation using Fourier lenses.

The search algorithm allows one to find elements in a list of N items fulfilling a given predicate in time $O(N^{1/2})$. We assume that the predicate f is given by a quantum circuit V_f and, as usual in this setting, we count the invocations of V_f (the "oracle"). It is straightforward to construct from V_f an operator S_f which flips the amplitudes of the states that fulfill the predicate, i.e. $S_f : |x\rangle \mapsto (-1)^{f(x)}|x\rangle$ for all basis states $|x\rangle$. The Grover algorithm relies on an averaging method called *inversion about average*, which is described in the following.

Consider the matrix

$$D_n := \begin{pmatrix} -1 + \frac{2}{2^n} & \frac{2}{2^n} & \cdots & \frac{2}{2^n} \\ \frac{2}{2^n} & -1 + \frac{2}{2^n} & \cdots & \frac{2}{2^n} \\ \vdots & \vdots & \ddots & \vdots \\ \frac{2}{2^n} & \frac{2}{2^n} & \cdots & -1 + \frac{2}{2^n} \end{pmatrix}. \tag{4.6}$$

D_n is a circulant matrix [214], i.e. $D_n = \text{circ}_G(-1+(2/2^n), (2/2^n), \ldots, (2/2^n))$ for any choice of a finite group G. To see this, we recall the definition of a general group circulant,

$$\text{circ}_G(v) := (v_{g_i^{-1} \cdot g_j})_{1 \leq i,j \leq |G|}$$

for a fixed ordering $G = \{g_1, \ldots, g_{|G|}\}$ of the elements of G and for a vector v which is labeled by G. To implement (4.6) on a quantum computer we apply the circulant for the group $\mathbf{Z}_2^n$, making use of the following theorem.

Theorem 4.12. *The Fourier transform* DFT_A *for a finite abelian group A implements a bijection between the set of A-circulant matrices and the set of diagonal matrices over $\mathcal{C}$. Explicitly, each circulant C is of the form*

$$C = \text{DFT}_A^{-1} \cdot \text{diag}(d_1, \ldots, d_n) \cdot \text{DFT}_A$$

and the vector $\mathbf{d} = (d_1, \ldots, d_n)$ *of diagonal entries is given by* $\mathbf{d} = \text{DFT}_A \cdot \mathbf{c}$, *where* $\mathbf{c}$ *is the first row of C.*

Fig. 4.9. (a) Equal distribution, (b) flip solutions, (c) inversion about average

Grover's original idea was to use correlations to amplify the amplitudes of the states fulfilling the predicate, i.e. to correlate – starting from an initial distribution, which is chosen to be the equal distribution $P(X = i) = 1/N, i = 1, \ldots, N$ – the vector $(-1, 1, \ldots, 1)$ with the probability distribution obtained by flipping the signs of the states fulfilling the predicate.

Starting from the equal distribution $H_2^{\otimes n}|0\rangle$, the amplification process in Grover's algorithm consists of an iterated application of the operator $-D_n S_f$ $O(\sqrt{2^n})$ times [215]. In Fig. 4.9, the steps of this procedure are illustrated in a qualitative way.

We present a realization of the Grover algorithm with a diffractive optical system. This is not a quantum mechanical realization in the sense of quantum computing, since such a system does not support a qubit architecture. However, the transition matrices are unitary and hence we can consider simulations of quantum algorithms via optical devices. These optical setups scale linearly with the dimension of the computational Hilbert space rather than with the logarithmic growth of a quantum register; nevertheless they have some remarkable properties, the best known of which is the ability to perform a Fourier transform by a simple application of a cylindrical lens [216]. This resembles the classical Fourier transform (corresponding to $\text{DFT}_{\mathbf{Z}_{2^n}}$ rather than $\text{DFT}_{\mathbf{Z}_2^n}$), which is correct because of the comments preceding Theorem 4.12.

Observe that by starting from this operation, we can easily perform correlations since this corresponds to a multiplication with a circulant matrix. Every circulant matrix can be realized optically using a so-called $4f$ setup, which corresponds to a factorization of a circulant matrix C into diagonal matrices and Fourier transforms (following Theorem 4.12): $C = \text{DFT}^{-1} \cdot D \cdot \text{DFT}$, where D is a given diagonal matrix (see also Fig. 4.10).

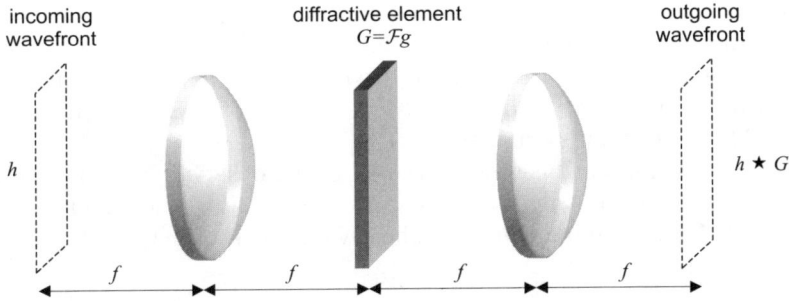

Fig. 4.10. Optical $4f$ setup that computes a convolution of h and g

Remark 4.3. The question of what linear transforms can be performed optically is equivalent to the question of what matrices can be factored into diagonal matrices and Fourier transforms, which correspond to diagonal matrices and circulant matrices. It has been shown in [217] that for fields $K \not\in \{GF(3), GF(5)\}$, every square matrix M with entries in K can be written as a product of circulant and diagonal matrices with entries in K. Furthermore, if M is unitary the circulant and diagonal factors can also be chosen to be unitary [217].

Remark 4.4. There have been several generalizations and modifications of the original Grover algorithm.

- The case of an unknown number of solutions fulfilling the given predicate is analyzed in [218].
- The issue of arbitrary initial distributions (instead of an equal distribution) is considered in [219].
- In [215] it is shown that instead of the diffusion operators D_n, almost any (except for a set of matrices of measure zero) unitary matrix can be used to perform the Grover algorithm.
- There is a quantum algorithm for the so-called collision problem which needs $O(\sqrt[3]{N/r})$ evaluations of a given r-to-one function f to find a pair of values which are mapped to the same element [220].
- The Grover algorithm has been shown to be optimal, i.e. the problem of finding an element in an unordered list takes $\Theta(\sqrt{N})$ operations on a quantum computer [200].
- Some other problems have been solved by a subroutine call to the Grover algorithm, e.g. a problem in communication complexity: the problem of deciding whether $A, B \subseteq \{1, \ldots, N\}$ are disjoint or not. Using the Grover algorithm, it can be shown that it is sufficient for the two parties, one holding A and the other holding B, to communicate $O(\sqrt{N} \log N)$ qubits to solve this problem [221].

4.5.2 Shor's Algorithm

In this section we briefly review Shor's factorization algorithm and show how the Fourier transform comes into play. It is known that factoring a number N is easy under the assumption that it is easy to determine the (multiplicative) order of an arbitrary element in $(\mathbf{Z}_N)^\times$. For a proof of this, we refer the reader to [222] and to Shor's original paper [9].

Once this reduction has been done, the following observation is the crucial step for the quantum algorithm. Let y be randomly chosen and let $\gcd(y, N) = 1$. To determine the multiplicative order r of $y \bmod N$, consider the function

$$f_y(x) := y^x \bmod N .$$

Clearly $f_y(x+r) = f_y(x)$, i.e. f_y is a periodic function with period r. The quantum algorithm to determine this period is as follows:

Algorithm 3 *Let N be given; determine $M = 2^m$ such that $N^2 \leq M \leq 2N^2$. This number M will be the length of the Fourier transform to be performed in the following.*

1. *Randomly choose y with $\gcd(y,N) = 1$.*
2. *Prepare the state $|0\rangle \otimes |0\rangle$ in two registers of lengths m and $\lceil \log_2 N \rceil$.*
3. *Application of the Hadamard transform $H_2^{\otimes m}$ to the left part of the register results in a superposition of all possible inputs*

$$\frac{1}{\sqrt{M}} \sum_{x=0}^{M-1} |x\rangle \otimes |0\rangle .$$

4. *We construct a unitary operation which computes the (partial function) $|x\rangle|0\rangle \mapsto |x\rangle|y^x \bmod N\rangle$ following Sect. 4.2.3 and 4.2.4. Calculation of $f_y(x) = y^x \bmod N$ yields for this superposition (normalization omitted)*

$$\sum_{x=0}^{M-1} |x\rangle \otimes |y^x \bmod N\rangle .$$

5. *Measuring the right part of the register gives a certain value z_0. The remaining state is the superposition of all x satisfying $f_y(x) = z_0$:*

$$\sum_{y^x = z_0} |x\rangle|z_0\rangle = \sum_{k=0}^{s-1} |x_0 + kr\rangle|z_0\rangle , \quad \text{where } y^{x_0} = z_0 \text{ and } s = \left\lfloor \frac{M}{r} \right\rfloor .$$

6. *Performing a QFT_M on the left part of the register leads to*

$$\sum_{l=0}^{M-1} \sum_{k=0}^{s-1} e^{2\pi \mathrm{i}(x_0 + kr)l/M} |l\rangle|z_0\rangle .$$

7. *Finally, a measurement of the left part of the register gives a value l_0.*

Application of this algorithm produces data from which the period r can be extracted after classical postprocessing involving Diophantine approximation (see Sect. 4.5.2).

A thorough analysis of this algorithm must take into account the overhead for the calculation of the function $f_y : x \mapsto y^x \bmod N$. However, after this function has been realized as a quantum network once (which can be obtained from a classical network for this function in polynomial time), the superposition principle applies since all inputs can be processed by *one* application of f_y.

The role played by Fourier transforms in this algorithm is twofold. In step 3 it is used to generate a superposition of all inputs from the ground state $|0\rangle$. This could have been done by *any* unitary transformation having an all-one vector in the first column. In step 6 we use the QFT_M to extract the information about the period r which was hidden in the graph of the function f_y.

Remark 4.5. It should be noted that for small numbers, as in the example of Fig. 4.11 and 4.12, an optical setup using Fourier lenses (cf. Fig. 4.10) could implement Shor's algorithm. This very example was initially simulated with the DIGIOPT® system [223].

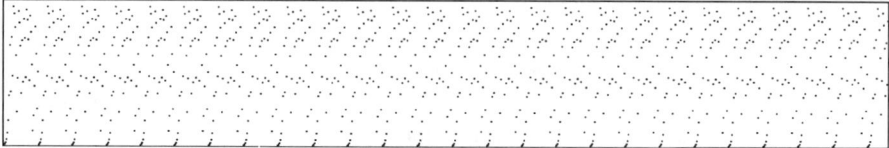

Fig. 4.11. Function graph of $f(x) = 2^x \mod 187$

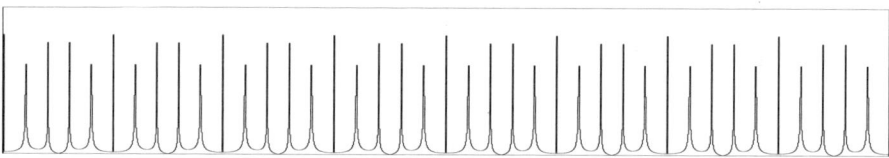

Fig. 4.12. The function shown in Fig. 4.11. transformed by a DFT of length 1024

An important question is the behavior of the states obtained after step 6 if the length M of the Fourier transform QFT_M we are using does not coincide with the period r of the function which is transformed. Note that this period r is exactly what has to be determined, and therefore the length N must be chosen appropriately in order to gain enough information about r from sampling. As it turns out, the choice $M = 2^m$ according to the condition $N^2 \leq M \leq 2N^2$ yields peaks in the Fourier domain which are sharply concentrated around the values lM/r [9].

In Fig. 4.11 and 4.12, this effect is illustrated for $N = 187$ and $M = 1024$. The choice of M does not fulfill the condition $N^2 \leq M \leq 2N^2$. Nevertheless, the characteristic peaks in the Fourier spectrum, which are sharply peaked around multiples of the inverse of the order, are apparent.

Let us now consider this state, which has been *oversampled* by transforming it with a Fourier transform of length M followed by a measurement, a little more closely. Constructive interference in $\sum_{k=0}^{s-1} e^{2\pi i(x_0+kr)l/M}|l\rangle$ will occur for those basis states $|l\rangle$ for which lr is close to M. The probability of measuring a specific l in this sum can be bounded by

$$\left|\frac{1}{\sqrt{sM}}\sum_{k=0}^{s-1}e^{2\pi ikrl/M}\right|^2 \geq \left|\frac{1}{\sqrt{sM}}\sum_{k=0}^{s-1}e^{\pi ir/M}\right|^2 = \frac{1}{sM}\frac{|1-e^{\pi irs/M}|^2}{|1-e^{\pi ir/M}|^2}$$

$$= \frac{1}{sM}\frac{|\sin[\pi rs/(2M)]|^2}{|\sin[\pi r/(2M)]|^2} \geq \frac{4}{\pi^2 r}.$$

The fractions l/M that we obtain from sampling fulfill the condition

$$\left|\frac{l}{M}-\frac{p}{r}\right|\leq\frac{1}{2M},$$

for some integer p. Because of the choice $M \geq N^2$ we obtain the result that l/M can be approximated efficiently by continued fractions, as described in the following section. This classical postprocessing completes the description of Shor's algorithm for finding the order r of y. To factor N we compute the least common divisors of $(y^{r/2}+1, N)$ and $(y^{r/2}-1, N)$, obtaining nontrivial factors of N if r is even and $y^{r/2} \not\equiv \pm 1 \mod N$.

A similar method can be applied to the discrete logarithm problem [9]. We mention that both the factoring and the discrete logarithm problem can be readily recognized as hidden-subgroup problems (see also Sect. 4.6.1): in the case of factoring, this corresponds to the group U generated by y and we are interested in the index $[\mathbf{Z}:U]$, which equals the multiplicative order of y. The discrete logarithm problem for $GF(q)^\times$ can be considered a hidden-subgroup problem for the group $G = \mathbf{Z} \times \mathbf{Z}$ and the function $f: G \to GF(q)^\times$ given by $f(x,y) \to \zeta^x \alpha^{-y}$, where ζ is the primitive element and α is the element for which we want to compute the logarithm.

The main difference from the situation in Simon's algorithm (see Sect. 4.3) is that in these cases we cannot apply the Fourier transform for the parent group G, since we do not know its order a priori. Rather, we have to compute larger Fourier transforms; preferably, the length is chosen to be a power of 2, to oversample. We then obtain the information from the sampled Fourier spectrum by classical postprocessing.

The basic features of the method of Fourier sampling which we invoked in Shor's factoring algorithm are also incorporated in Kitaev's algorithm for the abelian stabilizer problem [187]. At the very heart of Kitaev's approach is a method to measure the eigenvalues of a unitary operator U, supposing that the corresponding eigenvectors can be prepared. This estimation procedure becomes efficient if, besides U, the powers $U^{2^i}, i = 0, 1, \ldots$, can also be implemented efficiently [180, 187]. We must also mention that the method of Diophantine approximation is crucial for the phase estimation.

Diophantine Approximation In the following we briefly review some properties of the continued-fraction expansion of a real number. An important property a number can have is to be one of the convergents of such an expansion. More specifically, the following theorem holds, which is important in the sampling part of Shor's algorithm for recovering the exact eigenvalues of an operator from the data sampled.

Theorem 4.13. *For each $x \in \mathbb{R}$ and each fraction p/q fulfilling*

$$\left| x - \frac{p}{q} \right| < \frac{1}{2q^2} ,$$

the following holds: p/q is a convergent of the continued-fraction expansion of the real number x.

A proof of this theorem can be found in standard texts on elementary number theory (e.g. [224]). There exists a simple algorithm for producing an expansion of a rational number into a continued fraction:

Algorithm 4 *Let $x \in \mathbb{Q}$. Define $a_0 := \lfloor x \rfloor$, $x_1 := 1/(x-a_0)$, $a_1 := \lfloor x_1 \rfloor$, $x_2 := 1/(x_1 - a_1)$ and so on, until we obtain $x_i = a_i$ for the first time. Then we can write x as*

$$x = a_0 + \cfrac{1}{a_1 + \cfrac{1}{a_2 + \cfrac{1}{\cdots + \frac{1}{a_n}}}} .$$

We mention in addition that Algorithm 4 not only yields optimal approximations (if applied to elements $x \in \mathbb{R}$) but is also very efficient, since in principle just a Euclidean algorithm is performed.

4.5.3 Taxonomy of Quantum Algorithms

The quantum algorithms which have been discovered by now fall into two categories, the principles of which we shall describe in the following.

Entanglement-Driven Algorithms Suppose we are given a function $f : X \to Y$ from a (finite) domain X to a codomain Y. This function does not have to be injective; however, for a quantum computer to be able to perform f with respect to a suitably encoded X and Y, the function f has to be embedded into a unitary matrix V_f (cf. Sect. 4.2.3). We then can compute simultaneously the images of all inputs $x \in X$ using $|x\rangle|0\rangle \mapsto |x\rangle|f(x)\rangle$ for all $x \in X$ by preparing an equal superposition $\sum_{x \in X} |x\rangle$ in the X register and the ground state $|0\rangle$ in the Y register first, and then applying the quantum circuit V_f to obtain $\sum_{x \in X} |x\rangle|f(x)\rangle$. This entangled state can then be written as

$$\sum_{y \in \mathrm{Im}(f)} \left(\sum_{x : f(x) = y} |x\rangle \right) |y\rangle ,$$

i.e. we obtain a separation of the preimages of f. Measuring the second register leaves us with one of these preimages.

Pars pro toto, we mention the algorithms of Shor, which have been described in Sect. 4.5.2. In this case the function f is given by $f(x) = a^x \bmod N$, where N is the number to be factored and a is a random element in $\mathbf{Z}_N$.

Superposition-Driven Algorithms The principle of this class of algorithms is to amplify to a certain extent the amplitudes of a set of "good" states, e.g. states which are specified by a predicate given in the form of a quantum circuit, and on the other hand to shrink the amplitudes of the "bad" states.

Pars pro toto, we mention the Grover algorithm for searching an unordered list. We gave an outline of this algorithm in Sect. 4.5.1 and presented an optical implementation using Fourier lenses.

4.6 Quantum Signal Transforms

Abelian Fourier transforms have been used extensively in the algorithms in the preceding sections. We now consider further classes of unitary transformations which admit efficient realizations in a computational model of quantum circuits. In Sect. 4.6.1 we introduce generalized Fourier transforms which yield unitary transformations parametrized by (nonabelian) finite groups. Classically, advanced methods have been developed for the study of fast Fourier transforms for *solvable* groups [171, 174, 225, 226]. While it is an open question whether efficient quantum Fourier transforms exist for all finite solvable groups, it is possible to give efficient circuits for special classes (see Sect. 4.6.1). In Sect. 4.6.2 we consider a class of real orthogonal transformations useful in classical signal processing, for which quantum circuits of polylogarithmic size exist.

4.6.1 Quantum Fourier Transforms: the General Case

In Sect. 4.4 we have encountered the special case of the discrete Fourier transform for abelian groups. This concept can be generalized to arbitrary finite groups, which leads to an interesting and well-studied topic for classical computers. We refer to [171, 174, 225, 226] as representatives of a vast number of publications. The reader not familiar with the standard notations concerning group representations is referred to these publications and to standard references such as [213, 227]. Following [228], we briefly present the terms and notations from representation theory which we are going to use, and recall the definition of Fourier transforms.

The Wedderburn Decomposition Let ϕ be a regular representation of the finite group G. Then a *Fourier transform* for G is any matrix A that decomposes ϕ into irreducible representations with the additional property that equivalent irreducibles in the corresponding decomposition are equal. Regular representations are not unique, since they depend on the ordering of the elements of G. Also, note that this definition says nothing about the choice of the irreducible representations of G, which in the nonabelian case are not unique.

On the level of algebras, the matrix A is a constructive realization of the algebra isomorphism

$$\mathcal{C}G \cong \bigoplus_i M_{d_i}(\mathcal{C}) \tag{4.7}$$

of $\mathcal{C}$-algebras, where $M_{d_i}(\mathcal{C})$ denotes the full matrix ring of $d_i \times d_i$ matrices with coefficients in $\mathcal{C}$. This decomposition is also known as *Wedderburn decomposition* of the group algebra $\mathcal{C}G$.

We remind the reader of the fact that the decomposition in (4.7) is quite familiar from the theory of error-avoiding quantum codes and noiseless subsystems [229–232].

As an example of a Fourier transform, let $G = \mathbf{Z}_n = \langle x \mid x^n = 1 \rangle$ be the cyclic group of order n with regular representation $\phi = 1_E \uparrow_T G$, $T = (x^0, x^1, \ldots, x^{n-1})$, and let ω_n be a primitive nth root of unity.[4] Now $\phi^A = \bigoplus_{i=0}^{n-1} \rho_i$, where $\rho_i : x \mapsto \omega_n^i$ and $A = \mathrm{DFT}_n = (1/\sqrt{n})[\omega_n^{ij} \mid i, j = 0 \ldots n-1]$ is the (unitary) discrete Fourier transform well known from signal processing.

If A is a Fourier transform for the group G, then any fast algorithm for the multiplication with A is called a *fast Fourier transform* for G. Of course, the term *fast* depends on the complexity model chosen. Since we are primarily interested in the realization of a fast Fourier transform on a quantum computer (QFT), we first have to use the complexity measure κ, as derived in Sect. 4.2.1.

Classically, a *fast* Fourier transform is given by a factorization of the decomposition matrix A into a product of sparse matrices[5] [171, 174, 225, 233]. For a solvable group G, this factorization can be obtained recursively using the following idea. First, a normal subgroup of prime index $(G : N) = p$ is chosen. Using transitivity of induction, $\phi = 1_E \uparrow G$ is written as $(1_E \uparrow N) \uparrow G$ (note that we have the freedom to choose the transversals appropriately). Then $1_E \uparrow N$, which again is a regular representation, is decomposed (by

---

[4] The induction of a representation ϕ of a subgroup $H \leq G$ with transversal $T = (t_1, \ldots, t_k)$ is defined by $(\phi \uparrow_T G)(g) := [\dot{\phi}(t_i g t_j^{-1}) \mid i, j = 1 \ldots n]$, where $\dot{\phi}(x) := \phi(x)$ for $x \in H$ or else is the zero matrix of the appropriate size.

[5] Note that in general, sparseness of a matrix does not imply low computational complexity with respect to the complexity measure κ.

4 Quantum Algorithms: Applicable Algebra and Quantum Physics   131

recursion), yielding a Fourier transform B for N. In the last step, A is derived from B using a recursion formula.

A Decomposition Algorithm Following [228], we explain this procedure in more detail by first presenting two essential theorems (without proof) and then stating the actual algorithm for deriving fast Fourier transforms for solvable groups. The special tensor structure of the recursion formula mentioned above will allow us to use this algorithm as a starting point to obtain fast *quantum* Fourier transforms in the case where G is a 2-group (i.e. $|G|$ is a power of 2).

First we need Clifford's theorem, which explains the relationship between the irreducible representations of G and those of a normal subgroup N of G. Recall that G acts on the representations of N via inner conjugation: given a representation ρ of N and $t \in G$ we define $\rho^t : n \mapsto \rho(tnt^{-1})$ for $n \in N$.

Theorem 4.14 (Clifford's Theorem). *Let $N \trianglelefteq G$ be a normal subgroup of prime index p with (cyclic) transversal $T = (t^0, t^1, \ldots, t^{(p-1)})$, and denote by $\lambda_i : t \mapsto \omega_p^i$, $i = 0, \ldots, p-1$, the p irreducible representations of G arising from G/N. Assume ρ is an irreducible representation of N. Then exactly one of the two following cases applies:*

1. *$\rho \cong \rho^t$ and ρ has p pairwise inequivalent extensions to G. If $\overline{\rho}$ is one of them, then all are given by $\lambda_i \cdot \overline{\rho}$, $i = 0, \ldots, p-1$.*
2. *$\rho \not\cong \rho^t$ and $\rho \uparrow_T G$ is irreducible. Furthermore, $(\rho \uparrow_T G) \downarrow N = \bigoplus_{i=0}^{p-1} \rho^{t^i}$ and*

$$(\lambda_i \cdot (\rho \uparrow_T G))^{D \otimes \mathbf{1}_d} = \rho \uparrow_T G, \quad D = \mathrm{diag}(1, \omega_p, \ldots, \omega_p^{(p-1)})^i .$$

The following theorem provides the recursion formula and was used earlier by Beth [171] to obtain fast Fourier transforms based on the tensor product as a parallel-processing model.

Theorem 4.15. *Let $N \trianglelefteq G$ be a normal subgroup of prime index p having a transversal $T = (t^0, t^1, \ldots, t^{(p-1)})$, and let ϕ be a representation of degree d of N. Suppose that A is a matrix decomposing ϕ into irreducibles, i.e. $\phi^A = \rho = \rho_1 \oplus \ldots \oplus \rho_k$, and that $\overline{\rho}$ is an extension of ρ to G. Then*

$$(\phi \uparrow_T G)^B = \bigoplus_{i=0}^{p-1} \lambda_i \cdot \overline{\rho} ,$$

where $\lambda_i : t \mapsto \omega_p^i$, $i = 0, \ldots, p-1$, are the p irreducible representations of G arising from the factor group G/N,

$$B = (\mathbf{1}_p \otimes A) \cdot D \cdot (\mathrm{DFT}_p \otimes \mathbf{1}_d) , \quad \text{and} \quad D = \bigoplus_{i=0}^{p-1} \overline{\rho}(t)^i .$$

If, in particular, $\overline{\rho}$ is a direct sum of irreducibles, then B is a decomposition matrix of $\phi \uparrow_T G$.

In the case of a cyclic group G the formula yields exactly the well-known Cooley–Tukey decomposition (see also Sect. 4.4.1 and [209]), in which D is usually called the *twiddle matrix*.

Assume that $N \trianglelefteq G$ is a normal subgroup of prime index p with Fourier transform A and decomposition $\phi^A = \rho = \bigoplus_{i=1}^{m} \rho_i$. We can reorder the ρ_i such that the first, say k, ρ_i have an extension $\overline{\rho}_i$ to G and the other ρ_i occur as sequences $\rho_i \oplus \rho_i^t \oplus \ldots \oplus \rho_i^{t^{(p-1)}}$ of inner conjugates (cf. Theorem 4.14; note that the irreducibles ρ_i, $\rho_i^{t^j}$ have the same multiplicity since ϕ is regular). In the first case the extension may be calculated by Minkwitz's formula [234]; in the latter case each sequence can be extended by $\rho_i \uparrow_T G$ (Theorem 4.14, case 2). We do not state Minkwitz's formula here, since we shall not need it in the special cases treated later on. Altogether, we obtain an extension $\overline{\rho}$ of ρ and can apply Theorem 4.15. The remaining task is to ensure that equivalent irreducibles in $\bigoplus_{i=1}^{p} \lambda_i \cdot \overline{\rho}$ are equal. For summands of $\overline{\rho}$ of the form $\overline{\rho}_i$ we have the result that $\lambda_j \cdot \overline{\rho}_i$ and $\overline{\rho}_i$ are inequivalent, and hence there is nothing to do. For summands of $\overline{\rho}$ of the form $\rho_i \uparrow_T G$, we conjugate $\lambda_j \cdot (\rho_i \uparrow_T G)$ onto $\rho_i \uparrow_T G$ using Theorem 4.14, case 2.

Now we are ready to formulate the recursive algorithm for constructing a fast Fourier transform for a solvable group G due to Püschel et al. [228].

Algorithm 5 *Let $N \trianglelefteq G$ be a normal subgroup of prime index p with transversal $T = (t^0, t^1, \ldots, t^{(p-1)})$. Suppose that ϕ is a regular representation of N with (fast) Fourier transform A, i.e. $\phi^A = \rho_1 \oplus \ldots \oplus \rho_k$, fulfilling $\rho_i \cong \rho_j \Rightarrow \rho_i = \rho_j$. A Fourier transform B of G with respect to the regular representation $\phi \uparrow_T G$ can be obtained as follows.*

1. *Determine a permutation matrix P that rearranges the ρ_i, $i = 1, \ldots, k$, such that the extensible ρ_i (i.e. those satisfying $\rho_i = \rho_i^t$) come first, followed by the other representations ordered into sequences of length p equivalent to $\rho_i, \rho_i^t, \ldots, \rho_i^{t^{(p-1)}}$. (Note that these sequences need to be equal to $\rho_i, \rho_i^t, \ldots, \rho_i^{t^{(p-1)}}$, which is established in the next step.)*
2. *Calculate a matrix M which is the identity on the extensibles and conjugates the sequences of length p to make them equal to $\rho_i, \rho_i^t, \ldots, \rho_i^{t^{(p-1)}}$.*
3. *Note that $A \cdot P \cdot M$ is a decomposition matrix for ϕ, too, and let $\rho = \phi^{A \cdot P \cdot M}$. Extend ρ to G summand-wise. For the extensible summands use Minkwitz's formula; the sequences $\rho_i, \rho_i^t, \ldots, \rho_i^{t^{(p-1)}}$ can be extended by $\rho_i \uparrow_T G$.*
4. *Evaluate $\overline{\rho}$ at t and build $D = \bigoplus_{i=0}^{p-1} \overline{\rho}(t)^i$.*
5. *Construct a block-diagonal matrix C with Theorem 4.14, case 2, conjugating $\bigoplus_{i=0}^{p-1} \lambda_i \cdot \overline{\rho}$ such that equivalent irreducibles are equal. C is the identity on the extended summands.*

   Result:

   $$B = (\mathbf{1}_p \otimes A \cdot P \cdot M) \cdot D \cdot (\mathrm{DFT}_p \otimes \mathbf{1}_{|N|}) \cdot C \qquad (4.8)$$

   is a fast Fourier transform for G.

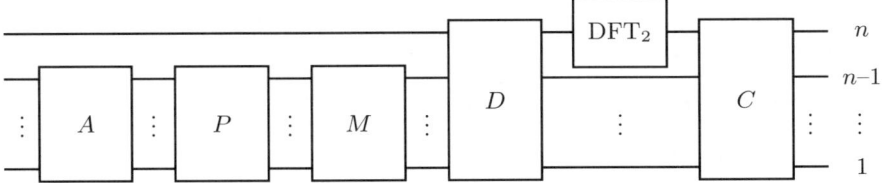

Fig. 4.13. Coarse quantum circuit visualizing Algorithm 5 for 2-groups

It is obviously possible to construct fast Fourier transforms on a classical computer for any solvable group by recursive use of this algorithm.

Since we are restricting ourselves to the case of a quantum computer consisting of qubits, i.e. two-level systems, we apply Algorithm 5 to obtain QFTs for 2-groups (i.e. $|G| = 2^n$ and thus $p = 2$). In this case the two tensor products occuring in (4.8) fit very well and yield a coarse factorization, as shown in Fig. 4.13. The lines in the figure correspond to the qubits as in Sect. 4.2.1, and a box ranging over more than one line denotes a matrix admitting no a priori factorization into a tensor product.

The remaining problem is the realization of the matrices A, P, M, D, C in terms of elementary building blocks as presented in Sect. 4.2.1. At present, the realization of these matrices remains a creative process to be performed for given (classes of) finite groups.

In [228] Algorithm 5 is applied to a class of nonabelian 2-groups, namely the 2-groups which contain a cyclic normal subgroup of index 2. These have been classified (see [235], Sect. 14.9, pp. 90–91), and for $n \geq 3$ there are exactly the following four isomorphism types:

1. the dihedral group $D_{2^{n+1}} = \langle x, y \mid x^{2^n} = y^2 = 1, \ x^y = x^{-1} \rangle$
2. the quaternion group $Q_{2^{n+1}} = \langle x, y \mid x^{2^n} = y^4 = 1, \ x^y = x^{-1} \rangle$
3. the group $QP_{2^{n+1}} = \langle x, y \mid x^{2^n} = y^2 = 1, \ x^y = x^{2^{n-1}+1} \rangle$
4. the quasi-dihedral group $QD_{2^{n+1}} = \langle x, y \mid x^{2^n} = y^2 = 1, \ x^y = x^{2^{n-1}-1} \rangle$.

Observe that the extensions 1, 3 and 4 of the cyclic subgroup $\mathbf{Z}_{2^n} = \langle x \rangle$ split, i.e. the groups have the structure of a semidirect product of $\mathbf{Z}_{2^n}$ with $\mathbf{Z}_2$. The three isomorphism types correspond to the three different embeddings of $\mathbf{Z}_2 = \langle y \rangle$ into $(\mathbf{Z}_{2^n})^\times \cong \mathbf{Z}_2 \times \mathbf{Z}_{2^{n-2}}$.

In [228] quantum circuits with polylogarithmic gate complexity are given for the Fourier transforms for each of these groups. See also [236, 237] for quantum Fourier transforms for nonabelian groups.

An Example: Wreath Products In this section we recall the definition of wreath products in general (see also [235, 238]) and, as an example, give efficient quantum Fourier transforms for a certain family of wreath products.

Definition 4.4. *Let G be a group and $H \subseteq S_n$ be a subgroup of the symmetric group on n letters. The wreath product $G \wr H$ of G with H is the set*

$$\{(\varphi, h) : h \in H, \varphi : [1, \ldots, n] \to G\}$$

equipped with the multiplication

$$(\varphi_1, h_1) \cdot (\varphi_2, h_2) := (\psi, h_1 h_2) i ,$$

where ψ is the mapping given by $i \mapsto \varphi_1(i^{h_2})\varphi_2(i)$ for $i \in [1, \ldots, n]$.

In other words, the wreath product is isomorphic to a semidirect product of the so-called *base group* $N := G \times \ldots \times G$, which is the n-fold direct product of (independent) copies of G with H, in symbols $G \wr H = N \rtimes H$, where H operates via permutation of the direct factors of N. So we can think of the elements as n-tuples of elements from G together with a permutation τ, and multiplication is done componentwise after a suitable permutation of the first n factors:

$$(g_1, \ldots, g_n; \tau) \cdot (g'_1, \ldots, g'_n; \tau') = (g_{\tau'(1)} g'_1, \ldots, g_{\tau'(n)} g'_n; \tau\tau') .$$

In this section we show how to compute a Fourier transform for certain wreath products on a quantum computer. We show how the general recursive method to obtain fast Fourier transforms on a quantum computer described in [228] can be applied directly in the case of wreath products $A \wr \mathbf{Z}_2$, where A is an arbitrary abelian 2-group. The recursion of the algorithm follows the chain

$$A \wr \mathbf{Z}_2 \triangleright A \times A \triangleright E ,$$

where the second composition factor is the base group. We first want to determine the irreducible representations of $G := A \wr \mathbf{Z}_2$. Let G^* be the base group of G, i.e. $G^* = A \times A$. G^* is a normal subgroup of G of index 2. Denoting by $\mathcal{A} = \{\chi_1, \ldots, \chi_k\}$ the set of irreducible representations of A, recall that the irreducible representations of G^* are given by the set $\{\chi_i \otimes \chi_j : i, j = 1, \ldots, k\}$ of pairwise tensor products (e.g. Sect. 5.6 of [212]).

Since $G^* \triangleleft G$, the group G operates on the representations of G^* via inner conjugation. Because G is a semidirect product of G^* with $\mathbf{Z}_2$, we can write each element $g \in G$ as $g = (a_1, a_2; \tau)$ with $a_1, a_2 \in A$ and we conclude

$$(\chi_1 \otimes \chi_2)^g = (\chi_1^{a_1} \otimes \chi_2^{a_2})^\tau = (\chi_1 \otimes \chi_2)^\tau,$$

i.e. only the factor group $G/G^* = \mathbf{Z}_2$ operates via permutation of the tensor factors. The operation of τ is to map $\chi_1 \otimes \chi_2 \mapsto \chi_2 \otimes \chi_1$.

Therefore, it is easy to determine the inertia groups (see [171,235] for definitions) T_ρ of a representation ρ of G^*. We have to consider two cases:

(a) $\rho = \chi_i \otimes \chi_i$. Then $T_\rho = G$, since permutation of the factors leaves ρ invariant.

(b) $\rho = \chi_i \otimes \chi_j, i \neq j$. Here we have $T_\rho = G^*$.

The irreducible representations of G^* fulfilling (a) extend to representations of G, whereas the induction of a representation fulfilling (b) is irreducible. In this case the restriction of the induced representation to G^* is, by Clifford theory, equal to the direct sum $\chi_1 \otimes \chi_2 \oplus \chi_2 \otimes \chi_1$.

Example 4.7. We consider the special case $W_n := \mathbf{Z}_2^n \wr \mathbf{Z}_2$, for which the quantum Fourier transforms have an especially appealing form.

Applying the design principles for Fourier transforms described in this section (see also [171, 174, 226, 228, 249]), we obtain the circuits for DFT_{W_n} in a straightforward way. Once we have studied the extension/induction behavior of the irreducible representations of G^*, the recursive formula

$$(\mathbf{1}_2 \otimes \text{DFT}_{G^*}) \cdot \bigoplus_{t \in T} \Phi(t) \cdot (\text{DFT}_{\mathbf{Z}_2} \otimes \mathbf{1}_{|A|^2}) \tag{4.9}$$

provides a Fourier transform for G. Here $\Phi(t)$ denotes the extension (as a whole) of the regular representation of G^* to a representation of G [171, 226, 228]. In the case of W_n, the transform DFT_{G^*} is the Fourier transform for $\mathbf{Z}_2^{2n}$ and therefore a tensor product of $2n$ Hadamard matrices.

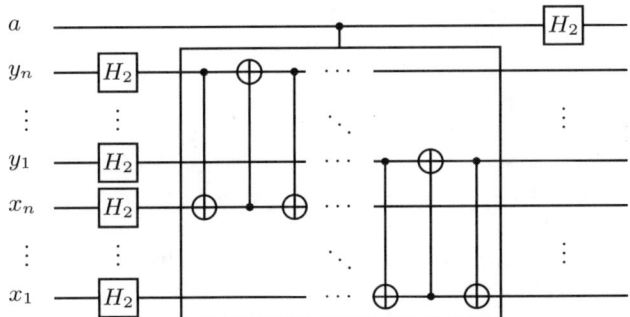

Fig. 4.14. The Fourier transform for the wreath product $\mathbf{Z}_2^n \wr \mathbf{Z}_2$

The circuits for the case of W_n are shown in Fig. 4.14. Obviously, the complexity cost of this circuit is linear in the number of qubits, since the conditional gate representing the evaluation at the transversal $\bigoplus_{t \in T} \Phi(t)$ can be realized with $3n$ Toffoli gates.

Nonabelian Hidden Subgroups We adopt the definition of the hidden-subgroup problem given in [239]. The history of the hidden subgroup problem parallels the history of quantum computing, since the algorithms of Simon [57] and Shor [9] can be formulated in the language of hidden subgroups (see, e.g.,

[240] for this reduction) for certain abelian groups. In [206] exact quantum algorithms (with a running time polynomial in the number of evaluations of the given black-box function and the classical postprocessing) are given for the hidden-subgroup problem in the abelian case. For the general case we use the following definition, which should be compared with Simon's problem in Sect. 4.3.

Definition 4.5 (The Hidden Subgroup Problem). *Let G be a finite group and $f : G \to R$ a mapping from G to an arbitrary domain R fulfilling the following conditions:*

(a) The function f is given as a quantum circuit, i.e. f can be evaluated in superpositions.
(b) There exists a subgroup $U \subseteq G$ such that f takes a constant value on each of the cosets gU for $g \in G$.
(c) Furthermore, f takes different values on different cosets.

The problem is to find generators for U.

We have already seen that Simon's algorithm and Shor's algorithms for the discrete logarithm and factoring can be seen as instances of abelian hidden-subgroup problems.

The hidden-subgroup problem for nonabelian groups provides a natural generalization of these quantum algorithms. Interesting problems can be formulated as hidden-subgroup problems for nonabelian groups, e.g. the graph isomorphism problem, which is equivalent to the problem of deciding whether a graph has a nontrivial automorphism group [241]. In this case G is the symmetric group S_n acting on a given graph Γ with n vertices. To reduce the graph automorphism problem to a hidden-subgroup problem for S_n, we encode Γ into a binary string in $R := \{0, 1\}^*$ and define f to be the mapping which assigns to a given permutation $\sigma \in S_n$ the graph Γ^σ. Progress in the direction of the graph isomorphism problem has been made in [242], but an efficient quantum algorithm solving the hidden-subgroup problem for S_n or the graph isomorphism problem is still not known.

In [239], the case of hidden subgroups of dihedral groups is addressed. The authors show that it is possible to solve the hidden-subgroup problem using only polynomially many queries to the black-box function f. However, the classical postprocessing takes exponential time in order to solve a nonlinear optimization problem. In [243] the hidden-subgroup problem for the wreath products $\mathbf{Z}_2^n \wr \mathbf{Z}_2$ is solved on a quantum computer, using polynomially many queries to f and efficient classical postprocessing which takes $O(n^3)$ steps. The fast quantum Fourier transforms for these groups, which have been derived in Sect. 4.6.1, have been used in this solution.

4.6.2 The Discrete Cosine Transform

In this section we address the problem of computing further signal transforms on a quantum computer. More specifically, we give a realization of the discrete cosine transforms (DCTs) of type II on a quantum computer, which is based on a well-known reduction to the computation of a discrete Fourier transform (DFT) of double length.

The DCT has numerous applications in classical signal processing, and hence, this transform might be useful in quantum computing and quantum state engineering as well. For efficient quantum signal transforms, and especially for wavelet transforms on quantum computers, see also [236, 244–246].

The discrete cosine transforms come in different flavors, varying slightly in their definitions. In this paper we restrict ourselves to the discrete cosine transform of type II (DCT_{II}) as defined below.

Recall that the DCT_{II} is the $N \times N$ matrix defined by (see [247], p. 11)

$$DCT_{II} := \left(\frac{2}{N}\right)^{1/2} \left(k_i \cos \frac{i(j+1/2)\pi}{N}\right)_{i,j=0,\ldots,N-1},$$

where $k_i = 1$ for $i = 1, \ldots, N-1$ and $k_0 = 1/\sqrt{2}$.

In [247] the DCT_{II} is shown to be a real and orthogonal (and hence unitary) transformation. Closely related is the discrete cosine transform DCT_{III}, which is defined as the transposed matrix (and hence the inverse) of DCT_{II}. Therefore, each efficient quantum circuit for the DCT_{III} yields one for the DCT_{II} and vice versa, since the inverse transform is obtained by reading the circuit backwards (where each elementary gate is conjugated and transposed).

Concerning the applications in classical signal processing, we should mention the well-known fact that the DCTs (of all families) are asymptotically equivalent to the Karhunen–Loève transform for signals produced by a first-order Markov process. The DCT_{II} is also used in the JPEG image compression standard [247].

For a given (normalized) vector $|x\rangle = (x(0), \ldots, x(N-1))^t$, we want to compute the matrix vector product $DCT_{II} \cdot |x\rangle$ on a quantum computer efficiently. Since DCT_{II} is a unitary matrix, it has a factorization into elementary quantum gates, and we seek a factorization of *polylogarithmic* length. Note that we restrict ourselves to the case $N = 2^n$ since matrices of this size fit naturally into the tensor product structure of the Hilbert space imposed by the qubits.

Theorem 4.16. *The discrete cosine transform* $DCT_{II}(2^n)$ *of length* 2^n *can be computed in* $O(n^2)$ *steps using one auxiliary qubit.*

Proof. The main idea (following Chap. 4 of [247]) is to reduce the computation of DCT_{II} to a computation of a DFT of double length.

Instead of the input vector $|x\rangle$ of length 2^n, we consider $|y\rangle$ of length 2^{n+1}, defined by

$$y(i) := \begin{cases} x(i)/\sqrt{2}, & i = 0, \ldots, 2^n - 1 \\ x(2^{n+1} - i - 1)/\sqrt{2}, & i = 2^n, \ldots, 2^{n+1} - 1 \end{cases}.$$

The input state of the quantum computer considered here, which has a register of length n carrying $|x\rangle$ and an extra qubit (which is initialized in the ground state), is $|\varphi_{in}\rangle = |0\rangle|x\rangle$ (written explicitly, this is $x(0)|00\ldots 0\rangle + \cdots + x(2^n)|01\ldots 1\rangle$), which is transformed by the circuit given in Fig. 4.15 to yield $|y\rangle$.

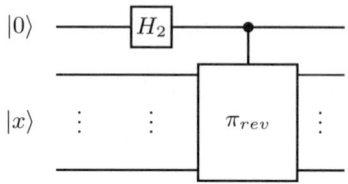

Fig. 4.15. Circuit that prepares the vector $|y\rangle$

As usual, H_2 is the Hadamard transformation and π_{rev} is the permutation obtained by performing a σ_x on each wire.

Application of $\text{DFT}_{2^{n+1}} := (1/\sqrt{2^{n+1}})[\omega_{2^{n+1}}^{i \cdot j}]_{i,j=0,\ldots,2^{n+1}-1}$, where $\omega_{2^{n+1}}$ denotes a primitive 2^{n+1}th root of unity, to the vector $|y\rangle$ yields the components

$$(\text{DFT}_{2^{n+1}} \cdot y)(m) = \frac{1}{\sqrt{2^{n+2}}} \sum_{i=0}^{2^n-1} (\omega_{2^{n+1}}^{m \cdot i} + \omega_{2^{n+1}}^{m \cdot (2^n - i - 1)}) x(i) ,$$

which holds for $m = 0, \ldots, 2^{n+1} - 1$.

Note that multiplication with $\text{DFT}_{2^{n+1}}$ can be performed in $O(n^2)$ steps on a quantum computer [9, 248], which is quite contrary to the classical FFT algorithm, which requires $O(n2^n)$ arithmetic operations. However, the tensor product recursion formula [171, 249] is well suited for direct translation into efficient quantum circuits [203].

Multiplication of the vector component $y(m)$ by the phase factor $\omega_{2^{n+1}}^{m/2}$ yields

$$\omega_{2^{n+1}}^{m/2}\left(\omega_{2^{n+1}}^{mi} + \omega_{2^{n+1}}^{m(2^{n+1}-i-1)}\right) = \omega_{2^{n+1}}^{m(i+1/2)} + \omega_{2^{n+1}}^{m(2^{n+1}-i-1/2)} ,$$

which means that this expression is equal to $2\cos[m(i+1/2)\pi/2^n]$. The multiplication with these (relative) phase factors corresponds to a diagonal matrix $T = \mathbf{1}_2 \otimes \text{diag}(\omega_{2^{n+1}}^{m/2}, m = 0, \ldots, 2^n - 1)$, which can be implemented by a tensor product $T = \mathbf{1}_2 \otimes T_n \otimes \cdots \otimes T_1$ of local operations, where $T_i = \text{diag}(1, \omega_{2^{n+2}}^{2^{i-1}})$ for $i = 1, \ldots, n$.

4 Quantum Algorithms: Applicable Algebra and Quantum Physics

Looking at the state obtained so far, we see that $|\varphi_{in}\rangle$ has been mapped to $|\varphi'\rangle$, the lower 2^n components of which are given by

$$\varphi'(m) = \frac{1}{\sqrt{2^n}} \sum_{i=0}^{2^n-1} \cos[m(i+1/2)\pi/2^n] x(i) ,$$

where $m = 0, \ldots, 2^{n+1} - 1$. Using an elementary property of the cosine, we see that the following holds for the components of this vector:

$$\varphi'(m) = -\varphi'(2^{n+1} - m), \text{ for } m = 1, \ldots, 2^n - 1 ,$$

and, furthermore, $\varphi'(0) = \sum_{i=0}^{2^n-1} x(i)$ and $\varphi'(2^n) = 0$. Hence we are nearly finished, since, except for cleaning up the help register, the x register has been transformed according to DCT_{II}.

Cleaning up the help register can be accomplished by the matrix

$$\begin{pmatrix} 1 & & & & & & -\frac{1}{\sqrt{2}} & & \\ & 1 & & & & & & & \\ & & \ddots & & & & & & \\ & & & \frac{1}{\sqrt{2}} & -\frac{1}{\sqrt{2}} & & & & \\ & & & 1 & & & & & \\ & & & \frac{1}{\sqrt{2}} & \frac{1}{\sqrt{2}} & & & & \\ & & & & & \ddots & & & \\ & & & & & & 1 & & \\ \frac{1}{\sqrt{2}} & & & & & & & \frac{1}{\sqrt{2}} \end{pmatrix} = \pi^{-1} \begin{pmatrix} 1 & & & & & & \\ & 1 & & & & & \\ & & \frac{1}{\sqrt{2}} & -\frac{1}{\sqrt{2}} & & & \\ & & \frac{1}{\sqrt{2}} & \frac{1}{\sqrt{2}} & & & \\ & & & & \ddots & & \\ & & & & & \frac{1}{\sqrt{2}} & -\frac{1}{\sqrt{2}} \\ & & & & & \frac{1}{\sqrt{2}} & \frac{1}{\sqrt{2}} \end{pmatrix} \pi ,$$

where the permutation matrix π arranges the columns $0, \ldots, 2^{n+1}$ in such a way that $(i, 2^{n+1} - i + 1)$ stand next to each other for $i = 1, \ldots, 2^n - 1$. This can be achieved by a quantum circuit, used also in [228], where this permutation appeared in the reordering of irreducible representations according to the action of a dihedral group. The circuit implementing π is given in Fig. 4.16 and can be performed in $O(n^2)$ operations. Here P_n is the cyclic shift $x \mapsto x + 1 \mod 2^n$ on the basis states, which can be implemented in $O(n^2)$ operations (see [228], Sect. 3).

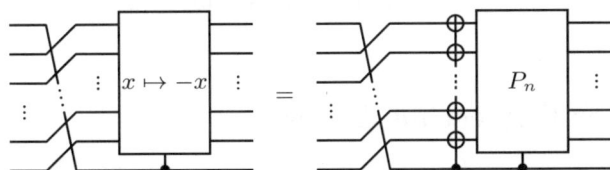

Fig. 4.16. Grouping the matrix entries pairwise via π

The matrix obtained after conjugation with π is a tensor product of the form $\mathbf{1}_{2^n} \otimes D_1$, where

$$D_1 := \frac{1}{\sqrt{2}} \begin{pmatrix} 1 & -1 \\ 1 & 1 \end{pmatrix} ,$$

up to a multiplication with the block diagonal matrix

$$\mathrm{diag}\Bigl(\frac{1}{\sqrt{2}}\begin{pmatrix} 1 & 1 \\ -1 & 1 \end{pmatrix}, \mathbf{1}_{2^{n+1}-2}\Bigr),$$

which in turn can be implemented by an $(n-1)$-fold controlled operation, i.e. in $O(n^2)$ elementary gates [172].

Overall, we obtain the circuit for the implementation of a DCT_{II} in $O(n^2)$ elementary gates shown in Fig. 4.17 (we have set $D_2 := D_1^{-1}$). □

We close this section with the remark that it is possible to derive other decompositions of DCT_{II} into a product of elementary quantum gates which have the advantage of being in-place, i.e. the overhead of one qubit that the realization given in the previous section needs can be saved [250].

This factorization of $\mathrm{DCT}_{II}(2^n)$ of length 2^n can also be computed in $O(n^2)$ elementary operations and does not make use of auxiliary registers.

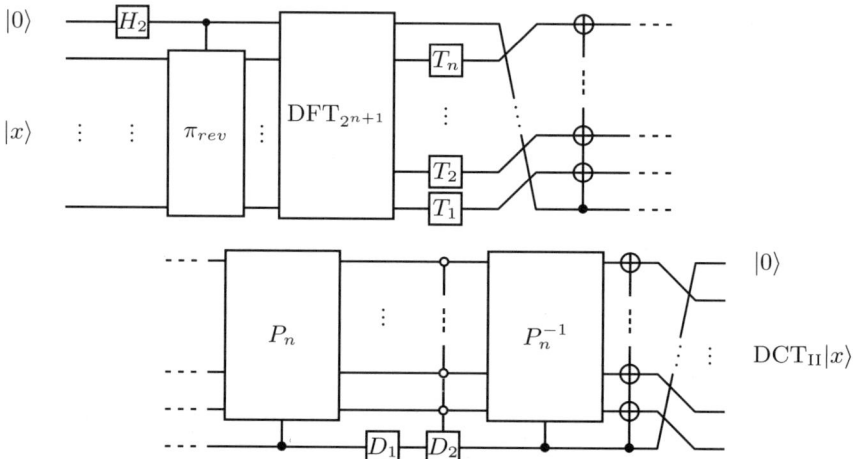

Fig. 4.17. Complete quantum circuit for DCT_{II} using one auxiliary qubit

4.7 Quantum Error-Correcting Codes

4.7.1 Introduction

The class of quantum error-correcting codes (QECCs) provides a master example that represents multiple aspects of the features of quantum algorithms. They are quantum algorithms per se, showing the applicability of the algorithmic concepts described so far to an intrinsically important area of quantum computation and quantum information theory, namely, the stabilization of quantum states.

This is achieved by methods which are based on signal-processing techniques at two levels:

- the Hilbert space of the quantum system itself
- the discrete vector space of the underlying combinatorial configurations.

The latter structure of finite-geometric spaces gives, furthermore, a deep insight into the features of entanglement, as the error-correcting codes known from classical coding theory share a particular feature, namely that of generating sets of codewords with maximal minimum distance by constructing suitable geometric configurations. On the other hand, these configurations nicely display the features of maximally entangled quantum states.

4.7.2 Background

We follow the presentation in [104], Chap. 13 for the background on the classical theory. A classical error-correcting code (ECC) consists of a set of binary words

$$\mathbf{x} = (x_0, \ldots, x_{n-1})$$

of length n, where each "bit" $x_i \in \{0, 1\}$ can take a value $x_i \in GF(2)$ in the finite field of two elements. With this elementary notion, the codewords can be treated as vectors of the n-dimensional $GF(2)$ vector space $GF(2)^n$, where the set of codewords is usually assumed to form a k-dimensional subspace $\mathcal{C} \le GF(2)^n$. This can be obtained canonically as the range $\text{im}(G)$ of a $GF(2)$ linear mapping $G : GF(2)^k \to GF(2)^n$ of the so-called encoder matrix, which maps k-bit messages onto n-bit codewords, adding a redundancy of $r = n - k$ bits in the r so-called parity check bits.

The characteristic parameters of such a linear code $\mathcal{C}$ are the rate $R = k/n$ and the minimum Hamming weight

$$w_H := \min\{\text{wgt}_H(\mathbf{c}) : \mathbf{c} \in \mathcal{C}, \mathbf{c} \ne 0\} \ .$$

The Hamming weight for a vector $\mathbf{x} = (x_0, \ldots, x_{n-1}) \in GF(2)^n$ is defined by $\text{wgt}_H(\mathbf{x}) := |\text{supp}(\mathbf{x})|$, where $\text{supp}(\mathbf{x}) := \{i \in [0, \ldots, n-1] : x_i \ne 0\}$.

It may be noted that the Hamming distance $d_H(\mathbf{u}, \mathbf{v}) := \#\{i \in [0, \ldots, n-1] : u_i \ne v_i\}$, which measures the number of bit flips necessary to change $\mathbf{v}$ into $\mathbf{u}$, is given by $d_H(\mathbf{u}, \mathbf{v}) = \text{wgt}_H(\mathbf{u} - \mathbf{v})$. Thus, since the code $\mathcal{C}$ is assumed to be a linear subspace of $GF(2)^n$, the equality for the minimum distance $d_H(\mathcal{C}) = \min_{\mathbf{u} \ne \mathbf{v}} d_H(\mathbf{u}, \mathbf{v}) = w_H$ is easily derived.

In constructing error-correcting codes, besides solving the parametric optimization problem of maximizing the rate of $\mathcal{C}$ and its minimum weight, it is a challenge to provide an efficient decoding algorithm at the same time. A decoder for $\mathcal{C}$ can in principle be built as follows:

Lemma 4.3. *The dual code $\mathcal{C}^\perp$ of $\mathcal{C}$, which is the $(n-k)$-dimensional subspace of $GF(2)^n$ orthogonal to $\mathcal{C}$ with respect to the canonical $GF(2)$ bilinear pairing (see also Sect. 4.4.2), is generated by any matrix $H : GF(2)^{n-k} \to GF(2)^n$ with $\mathrm{im}(H) = \mathcal{C}^\perp$.*

Obviously $\mathcal{C} = \mathrm{Ker}(H^t)$, the kernel of the transpose of H, and $G \cdot H^t = 0$.
□

For the sake of simplicity we assume that each codeword $\mathbf{c} \in \mathcal{C}$ is transmitted through a binary symmetric channel (BSC) (Fig. 4.18), which is the master model of a discrete memoryless channel [251]. A BSC is assumed to add the error vector $\mathbf{e} \in GF(2)^n$ independently of the codeword so that a noisy vector $\mathbf{u} = \mathbf{c} + \mathbf{e}$ is received with probability $q^{n-\mathrm{wgt}(\mathbf{e})} \cdot p^{\mathrm{wgt}(\mathbf{e})}$, where $q = 1 - p$.

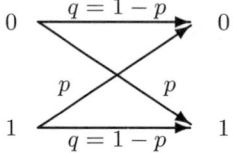

Fig. 4.18. Binary symmetric channel with parameter p

From the *syndrome*

$$\mathbf{s} := \mathbf{u} \cdot H^t = \mathbf{c} \cdot H^t + \mathbf{e} \cdot H^t = \mathbf{e} \cdot H^t$$

we see that the error pattern determines an affine subspace of $GF(2)^n$, namely a coset of $\mathcal{C}$ in $GF(2)^n$. In order to allow error correction, the syndrome has to be linked to the unknown error vector $\mathbf{e}$ uniquely, usually according to the maximum-likelihood decoding principle. Obviously, for the BSC this can be achieved by finding the unique codeword $\mathbf{c}$ that minimizes the Hamming distance to $\mathbf{u}$. For combinatorial reasons this is possible if

$$\mathrm{wgt}(\mathbf{e}) \leq \left\lfloor \frac{w_H - 1}{2} \right\rfloor.$$

Thus a code with minimum Hamming weight $w_H = 2t + 1$ is said to correct t errors per codeword.

4.7.3 A Classic Code

An example which today can be considered a classic in the field of science and technology is the application and design of (first-order) Reed–Müller codes $RM(1, m)$. This is not merely because they were a decisive piece of discrete mathematics in producing the first pictures from the surface of Mars in the early 1970s after the landing of Mariner 9 on 19 January 1972.

The description of their geometric construction and the method of error detection/correction of the corresponding wave functions, which we have taken from [104], Chap. 13, is quite similar to the concept of quantum error-correcting codes. In this section we describe Reed–Müller codes in a natural geometrical setting (cf. [103, 104, 252]).

In the *first-order Reed–Müller code* $RM(1, m)$, the codewords $\boldsymbol{f} \in GF(2)^{2^m}$ are $GF(2)$-linear combinations of class functions of the index-2 subgroups $H < \mathbf{Z}_2^m$ and their cosets. From this notion, a natural transform into the orthonormal basis of Walsh–Hadamard functions [253, 254] is given by the characters of $\mathbf{Z}_2^m$, i.e. the Hadamard transformation H_{2^m} (see also Sect. 4.2.1 and [255]). By these means, the $GF(2)$ vector

$$\boldsymbol{f} = (f(\boldsymbol{u}))_{\boldsymbol{u} \in \mathbf{Z}_2^m} \in GF(2)^{2^m}$$

is converted into the real (row) vector

$$\boldsymbol{F} = \left((-1)^{f(\boldsymbol{u})}\right)_{\boldsymbol{u} \in \mathbf{Z}_2^m}$$

by replacing an entry 0 in $\boldsymbol{f}$ by an entry 1 in $\boldsymbol{F}$ and an entry 1 in $\boldsymbol{f}$ by an entry -1 in $\boldsymbol{F}$. Modulation into a wavefunction $F(t)$ is then achieved by transmitting the step function in the interval $[0, 2^m - 1]$ defined by

$$F(t) = \sum_{i=0}^{2^m-1} F_{\text{Binary}(i)} 1_{[i,i+1]}(t) \ .$$

Note that for the first Hadamard coefficient $\widehat{F}_0$ of $F(t)$, which is given by

$$\widehat{F}_0 = \int_0^{2^m} F(t) \, \mathrm{d}t \ ,$$

the following identity holds:

$$\widehat{F}_0 = 2^m - 2\,\text{wgt}(\boldsymbol{f}) \ .$$

This provides a beautiful maximum-likelihood decoding device similar to that needed for quantum codes.

We shall illustrate this with the example of the first-order Reed–Müller code $RM(1, 3)$.

The codewords of the first-order Reed–Müller code $RM(1, m)$ with $m = 3$, of length 8, can be regarded as incidence vectors of special subsets of points of $AG(3, 2)$ (see Fig. 4.19). The following table of Boolean functions (see Sect. 4.2.2) $\boldsymbol{f}_i(x_3, x_2, x_1) = 1 \oplus x_i$ of incidence vectors,

$$\begin{aligned}
\mathbf{1} &= (1, 1, 1, 1, 1, 1, 1, 1) \text{ corresponding to the whole space}, \\
\boldsymbol{f}_1 &= (1, 0, 1, 0, 1, 0, 1, 0) \text{ corresponding to the 2-subspace } (*, *, 0), \\
\boldsymbol{f}_2 &= (1, 1, 0, 0, 1, 1, 0, 0) \text{ corresponding to the 2-subspace } (*, 0, *), \\
\boldsymbol{f}_3 &= (1, 1, 1, 1, 0, 0, 0, 0) \text{ corresponding to the 2-subspace } (0, *, *),
\end{aligned}$$

(4.10)

thus defines an $(m+1) \times 2^m$ generator matrix G. Its range is the following set of 16 codewords:

$(0,0,0,0,0,0,0,0)$ $(1,1,1,1,1,1,1,1)$ $(0,0,0,0,1,1,1,1)$ $(1,1,1,1,0,0,0,0)$
$(0,0,1,1,0,0,1,1)$ $(1,1,0,0,1,1,0,0)$ $(0,1,0,1,0,1,0,1)$ $(1,0,1,0,1,0,1,0)$
$(0,0,1,1,1,1,0,0)$ $(1,1,0,0,0,0,1,1)$ $(0,1,0,1,1,0,1,0)$ $(1,0,1,0,0,1,0,1)$
$(0,1,1,0,0,1,1,0)$ $(1,0,0,1,1,0,0,1)$ $(0,1,1,0,1,0,0,1)$ $(1,0,0,1,0,1,1,0)$.

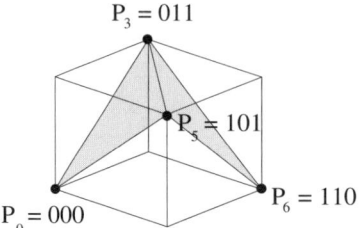

Fig. 4.19. Reed–Müller code $RM(1,3)$ interpreted as a 2-flat in $AG(3,2)$

Thus the subset $S = \{P_0, P_3, P_5, P_6\}$ determines the incidence vector $(1,0,0,1,0,1,1,0)$, which is the codeword $\boldsymbol{f}_1 + \boldsymbol{f}_2 + \boldsymbol{f}_3$ of $RM(1,3)$. The geometric interpretation of this codeword as an $(m-1)$-flat in $AG(3,2)$ is shown in Fig. 4.19.

With respect to G, the signal function $\boldsymbol{F}$ transmitted (see Fig. 4.20) thus corresponds to the $(m+1)$-bit input vector $(0,1,1,1)$.

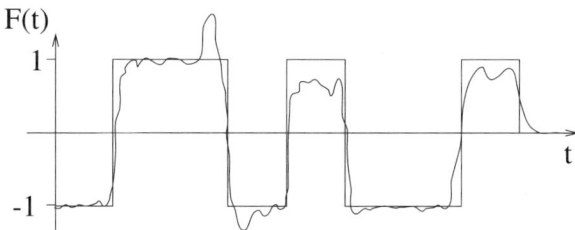

Fig. 4.20. Transmitted signal corresponding to the codeword $\boldsymbol{f}_0 + \boldsymbol{f}_1 + \boldsymbol{f}_2 = (1,0,0,1,0,1,1,0)$ with and without noise

In the more general case of Reed–Müller codes $RM(1,m)$ and their modulated signal functions, noise $e(t)$ is added to the signal function by the channel during transmission, so that the receiver will detect only a signal

$$\psi(t) = F(t) + e(t) \ .$$

The behavior of the $RM(1,m)$ demodulator–decoder device can be sketched on the basis of the underlying geometry as follows.

4 Quantum Algorithms: Applicable Algebra and Quantum Physics

Transformation of the received signal $\psi(t)$ into the orthonormal basis of Walsh–Hadamard functions $W_{\boldsymbol{u}}(t)$ of order m is achieved by computing the 2^m scalar products

$$\widehat{\psi}(\boldsymbol{u}) = \langle W_{\boldsymbol{u}} \mid \psi \rangle ,$$

where, for $\boldsymbol{u} \in GF(2)^m$, $W_{\boldsymbol{u}}(t)$ is given by

$$W_{\boldsymbol{u}}(t) = \sum_{j=1}^{2^m} (-1)^{\boldsymbol{u}\cdot \mathrm{Binary}(j-1)} \cdot 1_{[j-1,j]}(t) ,$$

corresponding to the uth row of the Hadamard matrix H_{2^n}.

As the received signal function is represented by a sampling vector $\boldsymbol{\psi} = \boldsymbol{F} + \boldsymbol{e}$, after the Walsh–Hadamard transformation one obtains $\widehat{\boldsymbol{\psi}} = \boldsymbol{\psi} \cdot H_{2^m} = \boldsymbol{F} H_{2^m} + \boldsymbol{e} H_{2^m}$. We estimate a maximum-likelihood signal function as follows. From $\widehat{\boldsymbol{F}} = \boldsymbol{F} \cdot H_{2^m}$, where

$$\widehat{F}(\boldsymbol{u}) = \sum_{\boldsymbol{v}} (-1)^{\boldsymbol{u}\cdot\boldsymbol{v}} F(\boldsymbol{v}) = \sum_{\boldsymbol{v}} (-1)^{\boldsymbol{u}\cdot\boldsymbol{v} + f(\boldsymbol{v}) \bmod 2} ,$$

we obtain the identity

$$|2^m - \widehat{F}(\boldsymbol{u})| = 2 \,\mathrm{wgt}(\boldsymbol{u}^\perp \oplus \boldsymbol{f}) ,$$

since $(\boldsymbol{u}\cdot\boldsymbol{v})_{\boldsymbol{v}\in GF(2)^m}$ is the incidence vector $\boldsymbol{u}^\perp$ of the hyperspace of $AG(2,m)$ orthogonal to the vector $\boldsymbol{u} \in GF(2)^m$. As the $RM(1,m)$ codes consist of all incidence vectors of such hyperspaces or their cosets (i.e. the complement), a minimum distance decoder is realized as shown in Fig. 4.21.[6]

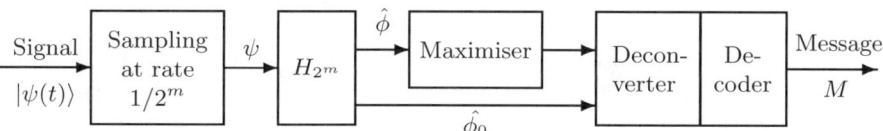

Fig. 4.21. Minimum-distance decoder for Reed–Müller codes (cf. [104], Chap. 13)

Here the maximizer computes

$$\boldsymbol{x} \in GF(2)^m \quad \text{such that} \quad |\psi(\boldsymbol{x})| = \max_{\boldsymbol{u} \in GF(2)^m} |\psi(\boldsymbol{u})| .$$

With the overall $\pm$ parity given by the sign $(\widehat{\psi}_0) = (-1)^{\varepsilon(\psi)}$, the deconverter produces a most likely codeword $\boldsymbol{f} = \varepsilon(\psi) \cdot \boldsymbol{1} + \boldsymbol{x}^\perp$. If the enumeration of

---
[6] With the fast Hadamard transform algorithm (see Sect. 4.2.1), this decoder requires $O(m 2^m)$ computational steps. Note that this is one of the earliest communication applications of generalized FFT algorithms (cf. [171]).

the generating hyperspaces u_i (cf. (4.10)) is chosen suitably, the decoder therefore reproduces the $(m+1)$-bit message $n = (\varepsilon, x_1, \ldots, x_m)$, which is a maximum-likelihood estimator of the word originally encoded into $f = M \cdot G$. The reader is urged to compare this decoding algorithm with the quantum decoding algorithm described in Sect. 4.7.4.

4.7.4 Quantum Channels and Codes

One of the basic features described so far in this contribution has been the use of quantum mechanical properties (entanglement, superposition) to speed up the solution of classical problems. Most notably, this basic idea of quantum computing has been present in the complex of hidden-subgroup problems (see Sect. 4.3, 4.5.2 and 4.6.1), where the problem was to identity an unknown subgroup of a given group out of exponentially many candidates, given a superposition over a coset of the unknown subgroup. In this section we shall take the dual point of view: we construct states which are simultaneous eigenstates of a suitably chosen subgroup of a fixed error group and have the additional property that an element of an unknown coset of this group – this models an error which happens to the states – can be identified and also corrected.

To construct these states, we rely on the powerful theory of classical ECC [184] introduced in the preceding sections. In what follows, we shall introduce the class of binary QECCs, the so-called CSS codes, referring to the elaborate article [255] by Beth and Grassl. These codes were independently discovered by Calderbank and Shor [82] and Steane [62].

Quantum Channels Before describing the construction of these codes, we shall loosely describe the similarities and differences between a classical binary symmetric channel (BSC) (see Sect. 4.7.2, Fig. 4.18) and a quantum channel (QC).

Much as in the idealized case of a BSC, where vectors $c \in GF(2)^n$ are transmitted, we shall consider the QC to be a carrier of kets $|\psi\rangle \in \mathcal{H}_{2^n} = \mathcal{C}^{2^n}$ spanned by the basis kets $|x\rangle$, where $x \in GF(2)^n$. In this system, so-called error operations, generated by local errors as bit-flip errors, and sign-flip errors can occur in superposition. Similarly to the case of a BSC, where the error group is isomorphic to $(GF(2)^n, \oplus) = \langle (e_1, \ldots, e_n) : e_i \in \{0,1\}, i = 1, \ldots, n \rangle$, in QC the error group

$$E = \langle e_1 \otimes \ldots \otimes e_n : e_i \in \{\mathrm{id}, \sigma_x, \sigma_z\}, i = 1, \ldots, n\} \rangle \,,$$

generated by the local bit flips and phase flips, represents all possible error operators in the quantum channel by the transition diagram described below.

Initially, the input wavefunction $|\psi\rangle$ will interact with the environment via $|\psi\rangle \mapsto |\psi\rangle|\epsilon\rangle$ through a modulator; within the channel, this waveform evolves under the error group according to the channel characteristics,

$$|\psi\rangle|\epsilon\rangle \xrightarrow[\text{error}]{\text{channel -}} \sum_{\gamma \in E}(\gamma|\psi\rangle)|\epsilon_\gamma\rangle , \qquad (4.11)$$

whereas, in the "environment", ancillae tacitly contain probability amplitudes for the occurrence of group elements.

Much as in the BSC, where only errors $\mathbf{e} \in GF(2)^n$ with a given maximal weight $\text{wgt}(\mathbf{e}) \leq t$ are allowed or assumed, in the QC the sum of the received ket

$$|\mu\rangle = \sum_{\substack{\gamma \in E \\ \text{wgt}(\gamma) \leq t}} \gamma|\psi\rangle|\epsilon_\gamma\rangle \qquad (4.12)$$

will range only over those group elements γ which are products of at most t local (single-bit) errors. Much in accordance with the classical case, for $\gamma = e_1 \otimes \ldots \otimes e_n \in E$ we define $\text{wgt}(\gamma)$ to be the number of occurrences of σ_x and σ_z needed to generate γ.

In order to protect quantum states against errors of this kind in a quantum channel, a so-called Pauli channel, a quantum error-correcting code must be constructed to map original quantum states into certain "protected" orthogonal subspaces, so that errors of the type of (4.12) cannot in practice damage the original state seriously. For this purpose, the theory of classical codes for the BSC can be successfully applied, as we describe below.

Quantum Codes The basic principle is to consider encoded quantum states which are superpositions of basis vectors belonging to classical codes, e.g.

$$|\mathcal{C}\rangle := \frac{1}{\sqrt{|\mathcal{C}|}} \sum_{\mathbf{c} \in \mathcal{C}} |\mathbf{c}\rangle .$$

First we note a surprising fact expressing an old result of error-correcting codes directly in the language of quantum theory.

Lemma 4.4 (MacWilliams). *Let $\mathcal{C} < GF(2)^n$ be an error-correcting code with its dual as usual. Let A be the elementary abelian group $(GF(2)^n, \oplus)$. Then DFT_A is the Hadamard transformation (see Sect. 4.4.2) $H_{2^n} := H_2 \otimes \ldots \otimes H_2$ (n factors). For each error vector $\mathbf{e} \in GF(2)^n$,*

$$H_{2^n}|\mathcal{C}\rangle = H_{2^n}\left(\frac{1}{\sqrt{|\mathcal{C}|}}\sum_{\mathbf{c} \in \mathcal{C}}|\mathbf{c} \oplus \mathbf{e}\rangle\right) = \frac{1}{\sqrt{|\mathcal{C}^\perp|}}\sum_{\mathbf{c} \in \mathcal{C}^\perp}(-1)^{\mathbf{c} \cdot \mathbf{e}}|\mathbf{c}\rangle . \qquad (4.13)$$

Proof. This result is due to the identity

$$\text{DFT}_A\left(\frac{1}{\sqrt{|U|}}\sum_{\mathbf{c} \in U}|\mathbf{c} \oplus \mathbf{e}\rangle\right) = \frac{1}{\sqrt{|U^\perp|}}\sum_{\mathbf{c} \in U^\perp}\beta(\mathbf{c}, \mathbf{e})|\mathbf{c}\rangle , \qquad (4.14)$$

which holds for all subgroups U of an abelian group A (see Sect. 4.4.2 and 4.4.3). □

The lemma says that any bit-flip error applied to the state $|\mathcal{C}\rangle$ will give a state whose support $\mathcal{C}^\perp$ is translation invariant, the shift being expressed only in the phases of the elements of $\mathcal{C}^\perp$. Dually, a phase-flip error in $|\mathcal{C}\rangle$ will occur as a bit-flip error in $|\mathcal{C}^\perp\rangle$.

From this, we deduce the following basic coding principle ([255], p. 462): given a classical binary linear (n, k) code $\mathcal{C}$ of length n and dimension k, the states of the related quantum code are given by

$$|\psi_{\boldsymbol{w}_i}\rangle = \frac{1}{\sqrt{|\mathcal{C}|}} \sum_{\boldsymbol{c} \in \mathcal{C}} (-1)^{\boldsymbol{c} \cdot \boldsymbol{w}_i} |\boldsymbol{c}\rangle \,, \tag{4.15}$$

with suitable $\boldsymbol{w}_i \in GF(2)^n$ encoding the ith basis ket of the initial state to be protected. In addition to the choice of the classical code $\mathcal{C}$, for the construction of a binary quantum code a subset $\mathcal{W} \subseteq GF(2)^n/\mathcal{C}^\perp$ has to be given to define the vectors $\boldsymbol{w}_i$.

Example 4.8. Let $\mathcal{C} := \{(0,0,0), (1,1,1)\} \subseteq GF(2)^3$ be the dual of the Reed–Müller code $\mathcal{R} = RM(1,3)$ shown in Fig. 4.19. Here $\mathcal{W} := GF(2)^3/\mathcal{C}^\perp$ provides an appropriate choice,

$$\boldsymbol{w}_0 = (0,0,0), \ \boldsymbol{w}_1 = (1,1,1) \,,$$

for the following encoding:

$$\begin{aligned} |\mathbf{0}\rangle &\mapsto |\psi_{\boldsymbol{w}_0}\rangle = |0,0,0\rangle + |1,1,1\rangle \,, \\ |\mathbf{1}\rangle &\mapsto |\psi_{\boldsymbol{w}_1}\rangle = |0,0,0\rangle - |1,1,1\rangle \,. \end{aligned} \tag{4.16}$$

Note that the state $|0,0,0\rangle + |1,1,1\rangle$ is the maximally entangled GHZ state. We remark that the GHZ state is, up to local unitary transformations, the unique maximally entangled state [256]. The "protected" subspace of code vectors is, by definition,

$$\mathcal{H}_0 = \{|\psi\rangle = \alpha|\psi_{\boldsymbol{w}_0}\rangle \beta|\psi_{\boldsymbol{w}_1}\rangle \mid |\alpha|^2 + |\beta|^2 = 1\} \,.$$

Since $\mathcal{C}$ is a one-error-correcting binary code, the quantum code is endowed with this property with respect to single bit-flip errors, i.e. it is protected against the error operators

$$\varepsilon_3 = id \otimes id \otimes \sigma_x, \ \varepsilon_2 = id \otimes \sigma_x \otimes id, \ \varepsilon_1 = \sigma_x \otimes id \otimes id \in \mathcal{U}(8) \,.$$

Obviously, the subspaces $\mathcal{H}_0$ and $\mathcal{H}_i = \varepsilon_i \mathcal{H}_0$ ($i = 1, 2, 3$) are mutually orthogonal, so that

$$\mathcal{H}^{2^3} = \bigoplus_{i=0}^{3} \mathcal{H}_i$$

is the direct sum of the four orthogonal spaces

$$\begin{aligned}
\mathcal{H}_0 &= \langle |0,0,0\rangle + |1,1,1\rangle, |0,0,0\rangle - |1,1,1\rangle \rangle\,, \\
\mathcal{H}_1 &= \langle |1,0,0\rangle + |0,1,1\rangle, |1,0,0\rangle - |0,1,1\rangle \rangle\,, \\
\mathcal{H}_2 &= \langle |0,1,0\rangle + |1,0,1\rangle, |0,1,0\rangle - |1,0,1\rangle \rangle\,, \\
\mathcal{H}_3 &= \langle |0,0,1\rangle + |1,1,0\rangle, |0,0,1\rangle - |1,1,0\rangle \rangle\,.
\end{aligned}$$

Thus, for any linear combination $\mathcal{E}$ of these single bit-flip errors ϵ_i, the encoded state $|\psi\rangle = \alpha|\psi_{\mathbf{w}_0}\rangle + \beta|\psi_{\mathbf{w}_1}\rangle$ is represented by

$$\mathcal{E}|\psi\rangle = \sum_{i=0}^{3} \lambda_i (\alpha \epsilon_i |\psi_{\mathbf{w}_0}\rangle + \beta \epsilon_i |\psi_{\mathbf{w}_1}\rangle)\,,$$

as a direct sum of four orthogonal vectors, each being "proportional" to $|\psi\rangle$. So, by a measurement, i.e. a random projection onto any of the spaces $\mathcal{H}_i$, or by applying a conditional gate $U = \mathrm{diag}(\epsilon_i^\dagger : i = 0, \ldots, 3)$, the original state can be reconstituted, thus correcting up to one bit-flip error.

But note that if, instead of $\mathcal{C}$, the code $\mathcal{R} = \mathcal{C}^\perp$ had been selected, this code construction could not have been successful, as $\mathcal{R} = RM(1,3)$ can detect one error but not correct it. It can be seen (see [255], p. 463) that the corresponding quantum code inherits this property of *detecting* one phase-error but not being capable of *correcting* it.

Motivated by and starting from this example and the properties and problems derived from it, we now quote the following theorem, which provides a method to obtain quantum codes from classical codes.

Theorem 4.17 (CSS Codes). *Suppose $\mathcal{C}_1, \mathcal{C}_2 \subseteq GF(2)^n$ are classical binary codes with parameters (n_1, k_1, d_1) and (n_2, k_2, d_2), respectively, fulfilling the additional requirement $\mathcal{C}_1^\perp \subseteq \mathcal{C}_2$. Let $W := \{\mathbf{w}_i : i = 1, \ldots, [\mathcal{C}_2 : \mathcal{C}_1^\perp]\}$ be a system of representatives of the cosets $\mathcal{C}_2^\perp / \mathcal{C}_1$. Then the set of states*

$$|\phi_{\mathbf{w}_i}\rangle := \frac{1}{\sqrt{|\mathcal{C}_1|}} \sum_{\mathbf{c} \in \mathcal{C}_1} (-1)^{\mathbf{c} \cdot \mathbf{w}_i} |\mathbf{c}\rangle$$

forms a quantum code $\mathcal{Q}$ which can correct $(d_1-1)/2$ bit errors and $(d_2-1)/2$ phase errors.

In practice, we have the following corollary in the case of weakly self-dual codes $\mathcal{C} \subseteq \mathcal{C}^\perp$, which were shown to be as good asymptotically in [82]:

Theorem 4.18. *Let $\mathcal{C}$ be a weakly self-dual binary code with dual distance d. Then the corresponding quantum code is capable of correcting up to $(d-1)/2$ errors.*

The construction in this theorem can be made more general, as Beth and Grassl have shown in [255]:

Theorem 4.19. *Let $\mathcal{C}^\perp$ be a weakly self-dual binary code. If for*

$$\mathcal{M}_0 := \{w \in GF(2)^n/\mathcal{C}^\perp \mid d_\mathrm{H}(\mathcal{C}^\perp, \mathcal{C}^\perp + w) \leq t\}$$

the following condition,

$$\forall w_i, w_j : i \neq j \Rightarrow \mathcal{M}_0 \cap (\mathcal{M}_0 + (w_i - w_j)) = \emptyset \,, \tag{4.17}$$

is satisfied, the quantum code can correct t errors, i.e. any error operator $\varepsilon \in E$ where the total number of positions exposed to bit flips or sign flips is at most t.

This leads to the following decoding algorithm.

Algorithm 6 (Quantum Decoding Algorithm) *Let $|\phi\rangle$ be encoded by a QECC according to Theorems 4.17–4.19. The received vector $\mathcal{E}|\phi\rangle$ will be decoded and reconstructed by the following steps:*

- *Perform a measurement to determine the bit-flip errors, i.e. project onto the code space $\mathcal{H}_\mathcal{C}$ or one of its orthogonal images $\mathcal{H}_{\mathcal{C}+e}$ under a bit-flip error e.*
- *Correct this bit-flip error ("subtract" e) by applying the corresponding tensor product of σ_x operators.*
- *Perform a Hadamard transformation.*
- *Perform a measurement to determine the sign-flip errors which corresponds to a bit-flip error in the actual bases.*
- *Correct this error.*
- *Reencode the final state.*

This decoding algorithm is easily understood from the point of view of binary codes, which have been designed as general constructions for the BSC (see Sect. 4.7.2). The reader should also observe the stunning analogy between this quantum decoder and the Green-machine decoder described in Sect. 4.7.3, Fig. 4.21.

4.8 Conclusions

We have presented an introduction to a computational model of quantum computers from a computer science point of view. Discrete Fourier transforms have been introduced as important subroutines used in several quantum algorithms. Throughout, unitary transformations which can be implemented in terms of elementary gates using a quantum circuit of polylogarithmic size have been of special interest; they yield an exponential speedup compared with the classical situation in many cases.

The quantum algorithms presented here exploit the fundamental principles of interference, superposition and entanglement that quantum physics offers. We have explored these principles in various algorithms, ranging from Shor's algorithms to algorithms for quantum error-correcting codes.

5 Mixed-State Entanglement and Quantum Communication

Michał Horodecki, Paweł Horodecki and Ryszard Horodecki

5.1 Introduction

Quantum entanglement is one of the most striking features of the quantum formalism [26]. It can be expressed as follows: *If two systems interacted in the past it is, in general, not possible to assign a single state vector to either of the two subsystems* [257]. This is what is sometimes called the principle of nonseparability. A common example of an entangled state is the singlet state [258],

$$\psi_- = \frac{1}{\sqrt{2}}(|01\rangle - |10\rangle) . \tag{5.1}$$

One can see that it cannot be represented as a product of individual vectors describing states of subsystems. Historically, entanglement was first recognized by Einstein, Podolsky and Rosen (EPR) [24] and by Schrödinger [5].[1] In their famous paper, EPR suggested a description of the world (called "local realism") which assigns an independent and objective reality to the physical properties of the well-separated subsystems of a compound system. Then EPR applied the criterion of local realism to predictions associated with an entangled state to conclude that quantum mechanics was incomplete. The EPR criticism was the source of many discussions concerning fundamental differences between the quantum and classical descriptions of nature.

The most significant progress toward the resolution of the EPR problem was made by Bell [25], who proved that local realism implies constraints on the predictions of spin correlations in the form of inequalities (called Bell's inequalities) which can be *violated* by quantum mechanical predictions for a system in the state (5.1). The latter feature of quantum mechanics, usually called *nonlocality*, is one of the most evident manifestations of quantum entanglement.

Information-theoretical aspects of entanglement were first considered by Schrödinger, who wrote [5], in the context of the EPR problem,
"Thus one disposes provisionally (until the entanglement is resolved by actual observation) of only a *common* description of the two in that space of higher dimension.

---
[1] In fact, entangled quantum states had been used in investigations of the properties of atomic and molecular systems [259].

This is the reason that knowledge of the individual systems can *decline* to the scantiest, even to zero, while that of the combined system remains continually maximal. Best possible knowledge of a whole does *not* include best possible knowledge of its parts – and that is what keeps coming back to haunt us".

In this way Schrödinger recognized a profoundly nonclassical relation between the information which an entangled state gives us about the whole system and the information which it gives us about the subsystems.

The recent development of quantum information theory has shown that entanglement can have important *practical* applications (see, e.g., [1–3, 260]). In particular, it turns out that entanglement can be used as a *resource* for communication of quantum states in an astonishing process called *quantum teleportation* [261].[2] In the latter, a quantum state is transmitted by use of a pair of particles in a singlet state (5.1) shared by the sender and receiver (usually referred to as Alice and Bob), and two bits of classical communication. However, in real conditions, owing to interaction with the environment, called decoherence, we encounter mixed states rather than pure ones. These mixed states can still possess some residual entanglement. More specifically, a mixed state is considered to be entangled if it is not a mixture of product states [263]. In mixed states the quantum correlations are weakened, and hence the manifestations of mixed-state entanglement can be very subtle [263–265]. Nevertheless, it appears that it can be used as a resource for quantum communication. Such a possibility is due to the discovery of *distillation* of entanglement [266]: by manipulation of noisy pairs, involving local operations and classical communication, Alice and Bob can obtain singlet pairs and apply teleportation. This procedure provides a powerful protection of the quantum data transmission against the environment.

Consequently, the fundamental problem was to investigate the structure of mixed-state entanglement, especially in the context of quantum communication. These investigations have led to discovery of *discontinuity* in the structure of mixed-state entanglement. It appears that there are at least two qualitatively different types of entanglement [71]: *free*, which is useful for quantum communication, and *bound*, which is a nondistillable, very weak and mysterious type of entanglement.

The present contribution is divided into two main parts. In the first part we report results of an investigation of the *mathematical* structure of entanglement. The main question is: given a mixed state, is it entangled or not? We present powerful tools that allow us to obtain the answer in many interesting cases. A crucial role is played here by the connection between entanglement and the theory of *positive maps* [267]. In contrast to *completely positive maps* [88], positive maps have not been applied in physics so far. The second part is devoted to the application of the entanglement of mixed states to quantum communication. Now, the leading question is: given an entangled state, can it be distilled? The mathematical tools worked out in the first part allow us

---

[2] For experimental realizations, see [262].

to answer the question. Surprisingly, the answer does not simplify the picture but, rather, reveals a new horizon including the basic question: what is the role of bound entanglement in nature?

Since entanglement is a basic ingredient of quantum information theory, the scope of application of the research presented here goes far beyond the quantum communication problem. The insight into the structure of entanglement of mixed states can be helpful in many subfields of quantum information theory, including quantum computing, quantum cryptography, etc.

Finally, it must be emphasized that our approach will be basically *qualitative*. Thus we shall not review here the beautiful work performed in the domain of *quantifying* entanglement [66, 268–271] (we shall only touch on this subject in the second part). Owing to the limited space for the present contribution, we shall also restrict our considerations to the entanglement of *bipartite* systems, even though a number of results have been recently obtained for *multipartite* systems (see, e.g., [272, 273]).

5.2 Entanglement of Mixed States: Characterization

We shall deal with states on the finite-dimensional Hilbert space $\mathcal{H}_{AB} = \mathcal{H}_A \otimes \mathcal{H}_B$. We shall call the system described by the Hilbert space $\mathcal{H}_{AB}$ the $n \otimes m$ system, where n and m are the dimensions of the spaces $\mathcal{H}_A$ and $\mathcal{H}_B$, respectively. An operator ϱ acting on $\mathcal{H}$ is a state if Tr $\varrho = 1$ and if it is a positive operator, i.e.

$$\text{Tr } \varrho P \geq 0 \tag{5.2}$$

for any projectors P (equivalently, positivity of an operator means that it is Hermitian and has nonnegative eigenvalues).

A state acting on the Hilbert space $\mathcal{H}_{AB}$ is called separable[3] if it can be approximated in the trace norm by states of the form

$$\varrho = \sum_{i=1}^{k} p_i \varrho_i \otimes \tilde{\varrho}_i , \tag{5.3}$$

where ϱ_i and $\tilde{\varrho}_i$ are states on $\mathcal{H}_A$ and $\mathcal{H}_B$, respectively. In finite dimensions one can use a simpler definition [274] (see also [269]): ϱ is separable if it is of the form (5.3) for some k (one can always find a $k \leq \dim \mathcal{H}_{AB}^2$). Note that the property of being entangled or not does not change if one subjects the state to a product unitary transformation $\varrho \to \varrho' = U_1 \otimes U_2 \varrho U_1^\dagger \otimes U_2^\dagger$. We call the states ϱ and ϱ' *equivalent*.

---

[3] The definition of separable states presented here is due to Werner [263], who called them classically correlated states.

We shall further need the following *maximally* entangled pure state of the $d \otimes d$ system:

$$\psi_+^d = \frac{1}{\sqrt{d}} \sum_{i=1}^{d} |i\rangle \otimes |i\rangle . \tag{5.4}$$

We shall denote the corresponding projector by P_+^d (the superscript d will usually be omitted). Then, for any state ϱ, the quantity $F = \langle \psi_+ | \varrho | \psi_+ \rangle$ is called the *singlet fraction*.[4] In general, by maximally entangled states we shall mean vectors ψ that are equivalent to ψ_+:

$$\psi = U_1 \otimes U_2 \psi_+ ,$$

where U_1, U_2 are unitary transformations. The most common two-qubit maximally entangled state is the singlet state (5.1). One can define the *fully entangled fraction* of a state ϱ of the $d \otimes d$ system by

$$\mathcal{F}(\varrho) = \max_{\psi} \langle \psi | \varrho | \psi \rangle , \tag{5.5}$$

where the maximum is taken over all maximally entangled vectors of the $d \otimes d$ system.

5.2.1 Pure States

If ϱ is a *pure* state, i.e. $\varrho = |\psi\rangle\langle\psi|$, then it is easy to check if it is entangled or not. Indeed, the above definition implies that it is separable if and only if $\psi = \psi_A \otimes \psi_B$, i.e. if either of its reduced density matrices is a pure state. Thus it suffices to find eigenvalues of either of the reductions. Equivalently, one can refer to the Schmidt decomposition [275] of the state. As one knows, for any pure state ψ there exist bases $\{e_i^A\}$, $\{e_i^B\}$ in the spaces $\mathcal{H}_A$ and $\mathcal{H}_B$ such that

$$\psi = \sum_{i=1}^{k} a_i |e_i^A\rangle \otimes |e_i^B\rangle, \quad k \leq \dim \mathcal{H}_{AB} , \tag{5.6}$$

where the positive coefficients a_i are called Schmidt coefficients. The state is entangled if at least two coefficients do not vanish. One finds that the positive eigenvalues of either of the reductions are equal to the squares of the Schmidt coefficients. In the next section we shall introduce a series of necessary conditions for the separability for mixed states. It turns out that all of them are equivalent to separability in the case of pure states [276, 277].

---

[4] In fact, the state ψ_+ used in the definition of the singlet fraction is a local transformation of the true singlet state. Nevertheless, we shall keep the name "singlet fraction", while using the state ψ_+, which is more convenient for technical reasons.

5.2.2 Some Necessary Conditions for Separability of Mixed States

A condition that is satisfied by separable states will be called a *separability criterion*. If a separability criterion is violated by the state, the state must be entangled. It is important to have strong separability criteria, i.e. those that are violated by the largest number of states if possible.

Since violation of the Bell inequalities is a manifestation of quantum entanglement, a natural separability criterion is constituted by the Bell inequalities. In [263] Werner first pointed out that separable states must satisfy all possible Bell inequalities.[5] The common Bell inequalities derived by Clauser, Horne, Shimony and Holt (CHSH) are given by [27]

$$\text{Tr } \varrho \mathcal{B} \leq 2 \, , \tag{5.7}$$

where the Bell–CHSH observable $\mathcal{B}$ is given by

$$\mathcal{B} = \hat{\boldsymbol{a}}\boldsymbol{\sigma} \otimes (\hat{\boldsymbol{b}} + \hat{\boldsymbol{b}}')\boldsymbol{\sigma} + \hat{\boldsymbol{a}}'\boldsymbol{\sigma} \otimes (\hat{\boldsymbol{b}} - \hat{\boldsymbol{b}}')\boldsymbol{\sigma} \, . \tag{5.8}$$

Here $\hat{\boldsymbol{a}}, \hat{\boldsymbol{a}}', \hat{\boldsymbol{b}}, \hat{\boldsymbol{b}}'$ are arbitrary unit vectors in $\mathbb{R}^3$, $\hat{\boldsymbol{a}}\boldsymbol{\sigma} = \sum_{i=1}^{3} a_i \sigma_i$, and σ_i are the Pauli matrices. For any given set of vectors we have a different inequality. In [278] we derived the condition for a two-qubit[6] state that was equivalent to satisfying all the inequalities jointly. This condition has the following form:

$$M(\varrho) \leq 1 \, , \tag{5.9}$$

where M is constructed in the following way. One considers the 3×3 real matrix T with entries $T_{ij} \equiv \text{Tr } \varrho \sigma_i \otimes \sigma_j$. Then M is equal to the sum of the two greater eigenvalues of the matrix $T^\dagger T$. This condition characterizes states violating the most common, and so far the strongest, Bell inequality for two qubits (see [279] in this context). While it is interesting from the point of view of nonlocality, it appears to be not a very strong separability criterion. Indeed, there exists [263] a large class of entangled states that satisfy all standard Bell inequalities.[7]

Another approach originated from the observation by Schrödinger [5] that an entangled state gives us more information about the total system than about subsystems. This gave rise to a series of entropic inequalities of the form [277, 280]

$$S(\varrho_A) \leq S(\varrho), \quad S(\varrho_B) \leq S(\varrho) \, , \tag{5.10}$$

---

[5] In [263] Werner also provided a very useful criterion based on the so-called *flip* operator (see Sect. 5.2.4).

[6] A *qubit* is the elementary unit of quantum information and denotes a two-level quantum system (i.e. a $2 \otimes 2$ system) [69].

[7] See [264, 265, 283] in the context of more sophisticated nonlocality criteria.

where $\varrho_A = \mathrm{Tr}_B\, \varrho$, and similarly for ϱ_B. The above inequalities were proved [277, 280–282] to be satisfied by separable states for four different entropies that are particular cases of the Renýi quantum entropies $S_\alpha = (1-\alpha)^{-1} \log \mathrm{Tr}\, \varrho^\alpha$:

$$S_0 = \log R(\varrho)\,,$$
$$S_1 = -\mathrm{Tr}\, \varrho \log \varrho\,,$$
$$S_2 = -\log \mathrm{Tr}\, \varrho^2\,,$$
$$S_\infty = -\log ||\varrho||\,, \qquad (5.11)$$

where $R(\varrho)$ denotes the rank of the state ϱ (the number of nonvanishing eigenvalues). The above inequalities are useful tools in many cases (as we shall see in Sect. 5.3.5, one of them allows us to obtain a bound on the possible rank of the *bound entangled* states); still, however, they are not very strong criteria.

A different approach, presented in [66], is based on local manipulations of entanglement (this approach was anticipated in [265]). The main idea is the following: a given state is entangled because parties sharing many systems (pairs of particles) in this state can produce a smaller number of pairs in a highly entangled state (of easily "detectable" entanglement) by local operations and classical communication (LOCC). This approach initiated a new field in quantum information theory: manipulating entanglement. The second part of this contribution will be devoted to this field. It also initiated the subject of the quantification of entanglement. Still, however, the seemingly simple qualitative question of whether a given state is entangled or not was not solved.

A breakthrough was achieved by Peres [284], who derived a surprisingly simple but very strong criterion. He noted that a separable state remains a positive operator if subjected to partial transposition (PT). We will call this the *positive partial transposition* (PPT) criterion.

To define partial transposition, we shall use the matrix elements of a state in some product basis:

$$\varrho_{m\mu,n\nu} = \langle m| \otimes \langle \mu| \varrho |n\rangle \otimes |\nu\rangle\,, \qquad (5.12)$$

where the kets with Latin and Greek letters form an orthonormal basis in the Hilbert space describing the first and the second system, respectively. Hence the partial transposition of ϱ is defined as

$$\varrho^{T_B}_{m\mu,n\nu} \equiv \varrho_{m\nu,n\mu}\,. \qquad (5.13)$$

The form of the operator ϱ^{T_B} depends on the choice of basis, but its eigenvalues do not. We shall say that a state "is PPT" if $\varrho^{T_B} \geq 0$; otherwise we shall say that the state "is NPT". The partial transposition is easy to perform in matrix notation. The state of the $m \otimes n$ system can be written as

$$\varrho = \begin{bmatrix} A_{11} & \ldots & A_{1m} \\ \ldots & \ldots & \ldots \\ A_{m1} & \ldots & A_{mm} \end{bmatrix}\,, \qquad (5.14)$$

with $n \times n$ matrices A_{ij} acting on the second ($\mathbb{C}^n$) space. These matrices are defined by their matrix elements $\{A_{ij}\}_{\mu\nu} \equiv \varrho_{i\nu,j\mu}$. Then the partial transposition can be realized simply by transposition (denoted by T) of all of these matrices, namely

$$\varrho^{T_B} = \begin{bmatrix} A_{11}^T & \dots & A_{1m}^T \\ \dots & \dots & \dots \\ A_{m1}^T & \dots & A_{mm}^T \end{bmatrix}. \tag{5.15}$$

Now, for any separable state ϱ, the operator ϱ^{T_B} must have still nonnegative eigenvalues [284]. Indeed, consider a partially transposed separable state

$$\varrho^{T_B} = \sum_i p_i \varrho_i \otimes (\tilde{\varrho}_i)^T. \tag{5.16}$$

Since the state $\tilde{\varrho}_i$ remains positive under transposition, so does the total state.

Note that what distinguishes the Peres criterion from the earlier ones is that it is *structural*. In other words, it does not say that some scalar function of a state satisfies some inequality, but it imposes constraints on the structure of the operator resulting from PT. Thus the criterion amounts to satisfying many inequalities at the same time. In the next section we shall see that there is also another crucial feature of the criterion: it involves a transposition that is a positive map but is *not* a completely positive one. This feature, abstracted from the Peres criterion, allows us to find an intimate connection between entanglement and the theory of positive maps.

Finally, it should be mentioned that necessary conditions for separability have been recently obtained in the infinite-dimensional case [285, 286]. In particular, the Peres criterion was expressed in terms of the Wigner representation and applied to Gaussian wave packets [286].

5.2.3 Entanglement and Theory of Positive Maps

To describe the very fruitful connection between entanglement and positive maps we shall need mathematical notions such as positive operators, positive maps and completely positive maps. In the following section we establish these notions. In the subsequent sections we use them to develop the characterization of the set of separable states.

Positive and Completely Positive Maps We start with the following notation. By $\mathcal{A}_A$ and $\mathcal{A}_B$ we shall denote the set of operators acting on $\mathcal{H}_A$ and $\mathcal{H}_B$, respectively. Recall that the set $\mathcal{A}$ of operators acting on some Hilbert space $\mathcal{H}$ constitutes a Hilbert space itself (a so-called Hilbert–Schmidt space) with a scalar product $\langle A, B \rangle = \text{Tr } A^\dagger B$. One can consider an orthonormal basis of operators in this space given by $\{|i\rangle\langle j|\}_{i,j=1}^{\dim \mathcal{H}}$, where $|i\rangle$ is a basis in the space $\mathcal{H}$. Since we are dealing with a finite dimension, $\mathcal{A}$ is in fact a space of matrices. Hence we shall sometimes denote it by M_d, where d is the dimension of $\mathcal{H}$.

The space of the linear maps from $\mathcal{A}_A$ to $\mathcal{A}_B$ is denoted by $\mathcal{L}(\mathcal{A}_A, \mathcal{A}_B)$. We say that a map $\Lambda \in \mathcal{L}(\mathcal{A}_A, \mathcal{A}_B)$ is positive if it maps positive operators in $\mathcal{A}_A$ into the set of positive operators, i.e. if $A \geq 0$ implies $\Lambda(A) \geq 0$. Finally, we need the definition of a completely positive (CP) map. We say [88] that a map $\Lambda \in \mathcal{L}(\mathcal{A}_A, \mathcal{A}_B)$ is completely positive if the induced map

$$\Lambda_n = \Lambda \otimes \mathbb{I}_n : \mathcal{A}_A \otimes \mathcal{M}_n \to \mathcal{A}_B \otimes \mathcal{M}_n \qquad (5.17)$$

is positive for all n; here $\mathbb{I}_n$ is the identity map on the space $\mathcal{M}_n$.[8] Thus the tensor product of a CP map and the identity maps positive operators into positive ones. An example of a CP map is $\varrho \to W\varrho W^\dagger$, where W is an arbitrary operator. As a matter of fact, the general form of a CP map is

$$\Lambda(\varrho) = \sum_i W_i \varrho W_i^\dagger \,. \qquad (5.18)$$

CP maps that do not increase the trace ($\operatorname{Tr} \Lambda(\varrho) \leq \operatorname{Tr} \varrho$) correspond to the most general physical operations allowed by quantum mechanics [88]. If $\operatorname{Tr} \Lambda(\varrho) = \operatorname{Tr} \varrho$ for any ϱ (we say the map is trace-preserving), then the operation can be performed with probability 1; otherwise, it can be performed with probability $p = \operatorname{Tr} \Lambda(\varrho)$.

It is remarkable that there are positive maps that are not CP: an example is just the transposition mentioned in the previous section. Indeed, if ϱ is positive, then so is ϱ^{T}, because

$$\operatorname{Tr} \varrho^{\mathrm{T}} P = \operatorname{Tr} \varrho P^{\mathrm{T}} \geq 0 \qquad (5.19)$$

and P^{T} is still some projector. We have used here the fact that $\operatorname{Tr} A^{\mathrm{T}} = \operatorname{Tr} A$. On the other hand, $\mathbb{I} \otimes T$ is no longer positive. One can easily check this, showing that $(\mathbb{I} \otimes T)P_+ \equiv P_+^{T_B}$ is not a positive operator.

A positive map is called *decomposable* [287] if it can be represented in the form

$$\Lambda = \Lambda_{\mathrm{CP}}^1 + \Lambda_{\mathrm{CP}}^2 \circ T \,, \qquad (5.20)$$

where Λ_{CP}^i are some CP maps. For low-dimensional systems ($\Lambda : M_2 \to M_2$ or $\Lambda : M_3 \to M_2$) the set of positive maps can be easily characterized. Namely, it has been shown [288, 289] that *all* the positive maps are decomposable in this case. If, instead, at least one of the spaces is $\mathcal{M}_n$ with $n \geq 4$, there exist nondecomposable positive maps [287, 289] (see the example in Sect. 5.2.4). No full characterization of positive maps has been worked out so far in this case.

---

[8] Of course, a completely positive map is also a positive one.

Characterization of Separable States via Positive Maps The fact that complete positivity is not equivalent to positivity is crucial for the problem of entanglement that we are discussing here. Indeed, trivially, the product states are mapped into positive operators by the tensor product of a positive map and an identity: $(\Lambda \otimes \mathbb{I})(\varrho \otimes \tilde\varrho) = (\Lambda\varrho) \otimes \tilde\varrho \geq 0$. Of course, the same holds for separable states. Then the main idea is that this property of separable states is essential, i.e., roughly speaking, if a state ϱ is entangled, then there exists a positive map Λ such that $(\Lambda \otimes \mathbb{I})\varrho$ is *not* positive. This means that one can seek the entangled states by means of the positive maps. Now the point is that not all of the positive maps can help us to determine whether a given state is entangled. In fact, the completely positive maps do not "feel" entanglement. Thus the problem of characterization of the set of the separable states reduces to the following: one should extract from the set of all positive maps some essential ones. As we shall see later, this is possible in some cases. Namely, it appears that for the $2 \otimes 2$ and $2 \otimes 3$ systems the transposition is the *only* such map. For higher-dimensional systems, apart from transposition, nondecomposable maps will also be relevant.

Consider the following lemma [267], which will lead us to the basic theorem relating entanglement and positive maps.

Lemma 5.1. *A state* $\varrho \in \mathcal{A}_A \otimes \mathcal{A}_B$ *is separable if and only if*

$$\mathrm{Tr}(A\varrho) \geq 0 \tag{5.21}$$

for any operator A satisfying $\mathrm{Tr}(AP \otimes Q) \geq 0$, *for all pure states P and Q acting on $\mathcal{H}_A$ and $\mathcal{H}_B$, respectively.*

Remark. Note that operator A, which is positive on product states (i.e. it satisfies $\mathrm{Tr} A\, P \otimes Q \geq 0$), is automatically Hermitian.

The lemma is a reflection of the fact that in real Euclidean space, a convex set and a point lying outside it can always be separated by a hyperplane.[9] Here, the convex set is the set of separable states, while the point is the entangled state. The hyperplane is determined by the operator A. Though this operator is not positive its restriction to product states is still positive. Thus, this operator has been called the "entanglement witness" [290], as it indicates the entanglement of some state (the first entanglement witness was provided in [263]; see Sect. 5.2.4). Now, to pass to positive maps, we shall use the isomorphism between entanglement witnesses and positive non-CP maps [291]. Note that, if we have any linear operator $A \in \mathcal{A}_A \otimes \mathcal{A}_B$, we can define a map

$$\langle k|\, \Lambda(|i\rangle\langle j|)\, |l\rangle = \langle i| \otimes \langle k|\, A\, |j\rangle \otimes |l\rangle\,, \tag{5.22}$$

---

[9] For infinite dimensions one must invoke the Hahn–Banach theorem, whose geometric form is a generalization of this fact.

which can be rephrased as follows:

$$\frac{1}{d}A = (\mathbb{I} \otimes \Lambda) P_+^d ,\qquad(5.23)$$

where $d = \dim \mathcal{H}_A$. Conversely, given a map, the above formula allows one to obtain a corresponding operator. It turns out that this formula also gives a one-to-one correspondence between entanglement witnesses and positive non-CP maps [291]. By applying this fact, one can prove [267] the following theorem:

Theorem 5.1. *Let ϱ act on the Hilbert space $\mathcal{H}_A \otimes \mathcal{H}_B$. Then ϱ is separable if and only if for any positive map $\Lambda : \mathcal{A}_B \to \mathcal{A}_A$ the operator $(\mathbb{I} \otimes \Lambda)\varrho$ is positive.*

As we mentioned, the relevant positive maps here are the ones that are not completely positive. Indeed, for the CP map Λ we have $(\mathbb{I} \otimes \Lambda)\varrho \geq 0$ for any state ϱ, and hence CP maps are of no use here. The above theorem presents, to our knowledge, the first application of the theory of positive maps in physics. So far, only completely positive maps have been of interest to physicists. As we shall see, the theorem has proved fruitful both for mathematics (the theory of positive maps) and for physics (the theory of entanglement).

Operational Characterization of Entanglement in Low Dimensions ($2 \otimes 2$ and $2 \otimes 3$ Systems) The first conclusion derived from the theorem is an operational characterization of the separable states in low dimensions ($2 \otimes 2$ and $2 \otimes 3$). This follows from the previously mentioned result that positive maps in low dimensions are decomposable. Then the condition $(\mathbb{I} \otimes \Lambda)\varrho \geq 0$ reads $(\mathbb{I} \otimes \Lambda_1^{\mathrm{CP}})\varrho + (\mathbb{I} \otimes \Lambda_2^{\mathrm{CP}})\varrho^{T_B}$. Now, since ϱ is positive and Λ_1^{CP} is CP, the first term is always positive. If ϱ^{T_B} is positive, then the second term is also positive, and hence their sum is a positive operator. Thus, to check whether for all positive maps we have $(\mathbb{I} \otimes \Lambda)\varrho \geq 0$, it suffices to check only transposition. One obtains the following [267] (see also [292]):

Theorem 5.2. *A state ϱ of a $2 \otimes 2$ or $2 \otimes 3$ system is separable if and only if its partial transposition is a positive operator.*

Remark. Equivalently, one can use the partial transposition with respect to the first space.

The above theorem is an important result, as it allows one to determine unambiguously whether a given quantum state of a $2 \otimes 2$ or a $2 \otimes 3$ system can be written as mixture of product states or not. The necessary and sufficient condition for separability here is surprisingly simple; hence it has found many applications. In particular, it has been applied in the context of broadcasting entanglement [293], quantum information flow in quantum copying networks

[294], disentangling machines [295], imperfect two-qubit gates [296], analysis of the volume of the set of entangled states [297, 298], decomposition of separable states into minimal ensembles or pseudo-ensembles [299], entanglement splitting [300] and analysis of entanglement measures [270, 301, 302].

In Sect. 5.3.2 we describe the first application [303]: by use of this theorem we show that any entangled two-qubit system can be distilled, and hence is useful for quantum communication.

**Higher Dimensions – Entangled States
with Positive Partial Transposition** Since the Størmer–Woronowicz characterization of positive maps applies only to low dimensions, it follows that for higher dimensions partial transposition will not constitute a necessary and sufficient condition for separability. Thus there exist states that are entangled but are PPT (see Fig. 5.1). The first explicit examples of an entangled but PPT state were provided in [274]. Later on, it became apparent, that the mathematical literature concerning nondecomposable maps contains examples of matrices that can be treated as prototypes of PPT entangled states [292, 304].

We shall now describe the method of obtaining such states presented in [274], as it has proved to be a fruitful direction in searching for PPT entangled states. Section 5.3 will describe the motivation for undertaking the very tedious task of this search – the states represent a curious type of entanglement, namely bound entanglement.

To find the desired examples we must find an entangled PPT state. Of course, we cannot use the strongest tool so far described, i.e. the PPT criterion, because we are actually trying to find a state that is PPT. So we must derive a criterion that would be *stronger* in some cases. It appears that the range[10] of the state can tell us much about its entanglement in some cases. This is contained in the following theorem, derived in [274] on the basis of the analogous condition for positive maps considered in [289].

Theorem 5.3. (Range Criterion). *If a state ϱ acting on the space $\mathcal{H}_{AB}$ is separable, then there exists a family of product vectors $\psi_i \otimes \phi_i$ such that*

(a) *they span the range of ϱ*
(b) *the vectors $\{\psi_i \otimes \phi_i^*\}_{i=1}^k$ span the range of ϱ^{T_B} (where $*$ denotes complex conjugation in the basis in which partial transposition was performed).*

In particular, any of the vectors $\psi_i \otimes \phi_i^*$ belongs to the range of ϱ.

Now, in [274] there were presented two examples of PPT states violating the above criterion. We shall present the example for a $2 \otimes 4$ case.[11] The

---
[10] The range of an operator A acting on the Hilbert space $\mathcal{H}$ is given by $R(A) = \{A(\psi) : \psi \in \mathcal{H}\}$. If A is a Hermitian operator, then the range is equivalent to the support, i.e. the space spanned by its eigenvectors with nonzero eigenvalues.
[11] This is based on an example concerning positive maps [289].

(a)

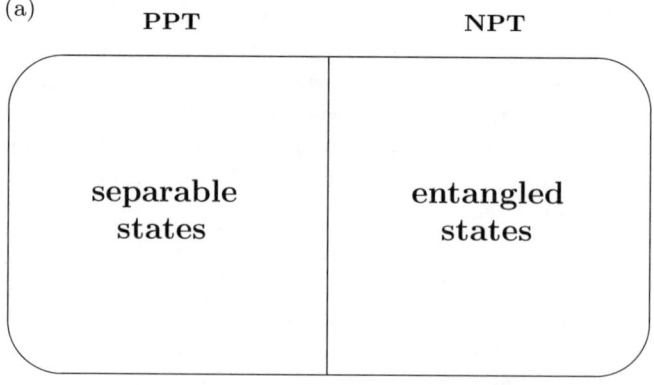

(b)

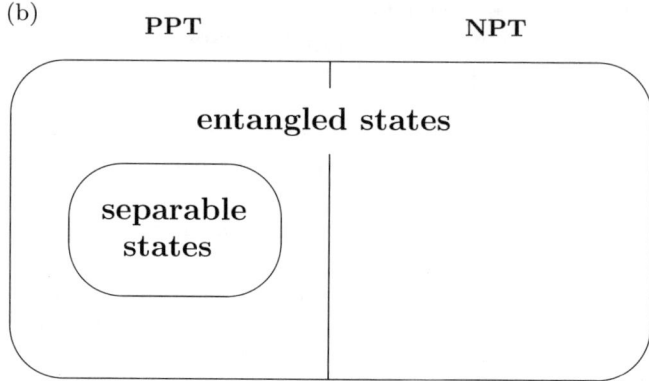

Fig. 5.1. Structure of entanglement of mixed states for $2\otimes 2$ and $2\otimes 3$ systems (**a**) and for higher dimensions (**b**)

matrix is written in the standard product basis $\{|ij\rangle\}$:

$$\varrho_b = \frac{1}{7b+1} \begin{bmatrix} b & 0 & 0 & 0 & 0 & b & 0 & 0 \\ 0 & b & 0 & 0 & 0 & 0 & b & 0 \\ 0 & 0 & b & 0 & 0 & 0 & 0 & b \\ 0 & 0 & 0 & b & 0 & 0 & 0 & 0 \\ 0 & 0 & 0 & 0 & \frac{1+b}{2} & 0 & 0 & \frac{\sqrt{1-b^2}}{2} \\ b & 0 & 0 & 0 & 0 & b & 0 & 0 \\ 0 & b & 0 & 0 & 0 & 0 & b & 0 \\ 0 & 0 & b & 0 & \frac{\sqrt{1-b^2}}{2} & 0 & 0 & \frac{1+b}{2} \end{bmatrix}, \qquad (5.24)$$

where $0 < b < 1$. Now, by performing PT as defined by (5.15), we can check that $\varrho_b^{T_B}$ remains a positive operator. By a tedious calculation, one can check that *none* of the product vectors belonging to the range of ϱ_b, if they are partially conjugated (as stated in the theorem), belongs to the range of $\varrho_b^{T_B}$. Thus the condition stated in the theorem is drastically violated, and hence the state is entangled. As we shall see further, the entanglement is masked so subtly that it cannot be distilled at all!

Range Criterion and Positive Nondecomposable Maps The separability criterion given by the above theorem has been fruitfully applied in the search for PPT entangled states [273, 305, 306]. Theorem 5.3 was applied in [307], where a technique of subtraction of product vectors from the range of the state was used to obtain the best separable approximation (BSA) of the state. As a tool, the authors considered subspaces containing *no* product vectors. Note that the (normalized) projector onto such a subspace must be entangled, as condition (a) of the theorem is not satisfied. This approach was successfully applied in [273] (see also [290, 305]) and, in connection with the seemingly completely different concept of *unextendible product bases*, produced an elegant, and so far the most transparent, method of construction of PPT entangled states.

To describe the construction,[12] one needs the following definition [273]:

Definition 5.1. *A set of product orthogonal vectors in $\mathcal{H}_{AB}$ that*

(a) has fewer elements than the dimension of the space
(b) is such that there does not exist any product vector orthogonal to all of them

is called an unextendible product basis (UPB).

Here we recall an example of such basis in the $3 \otimes 3$ system:

$$|v_0\rangle = \frac{1}{\sqrt{2}}|0\rangle|0-1\rangle, \quad |v_2\rangle = \frac{1}{\sqrt{2}}|2\rangle|1-2\rangle,$$

$$|v_1\rangle = \frac{1}{\sqrt{2}}|0-1\rangle|2\rangle, \quad |v_3\rangle = \frac{1}{\sqrt{2}}|1-2\rangle|0\rangle,$$

$$|v_4\rangle = \frac{1}{3}|0+1+2\rangle|0+1+2\rangle. \tag{5.25}$$

Of course, the above five vectors are orthogonal to each other. However, each of the two subsets $\{|v_0\rangle, |v_1\rangle, |v_4\rangle\}$ and $\{|v_2\rangle, |v_3\rangle, |v_4\rangle\}$ spans the full three-dimensional space. This prevents the existence of a sixth product vector that would be orthogonal to all five of them. How do we connect this with the problem we are dealing with in this section? The answer is: via the subspace

---

[12] The construction applies to the multipartite case [273]; in the present review we consider only bipartite systems.

complementary to the one spanned by these vectors. Indeed, suppose that $\{w_i = \phi_i \otimes \psi_i\}_{i=1}^k$ is a UPB. For a $d \otimes d$ system, consider the projector $P = \sum_{i=1}^k |w_i\rangle\langle w_i|$ onto the subspace $\mathcal{H}$ spanned by the vectors w_i (dim $\mathcal{H} = k$). Now, consider the state uniformly distributed on its orthogonal complement $\mathcal{H}_\perp$ (dim $\mathcal{H}_\perp = d^2 - k$),

$$\varrho = \frac{1}{d^2 - k}(\mathbb{I} - P) . \tag{5.26}$$

The range of the state ($\mathcal{H}_\perp$) contains no product vectors: otherwise one would be able to extend the product basis $\{w_i\}$. Then, by Theorem 5.3, the state must be entangled. Let us now calculate ϱ^{T_B}. Since $w_i = \phi_i \otimes \psi_i$, then $(|w_i\rangle\langle w_i|)^{T_B} = |\tilde{w}_i\rangle\langle \tilde{w}_i|$, where $\tilde{w}_i = \phi_i \otimes \psi_i^*$. The vectors $\tilde{w}_i$ are orthogonal to each other, so that the operator $P^{T_B} = \sum_i |\tilde{w}_i\rangle\langle \tilde{w}_i|$ is a projector. Consequently, the operator $(\mathbb{I} - P)^{T_B} = \mathbb{I} - P^{T_B}$ is also a projector, and hence it is positive. We conclude that ϱ is PPT.

A different way of obtaining examples of PPT entangled states can be inferred from the papers devoted to the search for nondecomposable positive maps in the mathematical literature [292, 304] (see Sect. 5.2.3). A way to find a nondecomposable map is the following. One constructs some map Λ and proves somehow that it is positive. Thus one can guess some (possibly unnormalized) state ϱ that is PPT. Now, if $(\mathbb{I} \otimes \Lambda)\varrho$ is not positive, then Λ cannot be decomposable, as shown in the discussion preceding Theorem 5.2. At the same time, the state must be entangled. In Sect. 5.2.4 we present an example of a PPT entangled state (based on [288]) found in this way. Thanks to its symmetric form, the state allowed researchers to reveal the first quantum effect produced by bound entanglement (see Sect. 5.3.7).

Thus a possible direction for exploring the "PPT region" of entanglement is to develop the description of nondecomposable maps. However, it appears that there can be also a "back-reaction": exploration of the PPT region may allow us to obtain new results on nondecomposable maps. It turns out that the UPB method described above allows for easy construction of new nondecomposable maps [308]. We direct the interested reader to the original article, as well as [290]. We note only that to find a nondecomposable map, one needs only to construct some UPB. Then the procedure is automatic, like the procedure described above. To our knowledge, this is the first *systematic* way of finding nondecomposable maps.

5.2.4 Examples

We present here a couple of examples, illustrating the results contained in previous sections. In particular, we introduce two families of states that play important roles in the problem of distillation of entanglement.

5 Mixed-State Entanglement and Quantum Communication

Reduction Criterion for Separability As mentioned in Sect. 5.2.3, if Λ is a positive map, then for separable states we have

$$(\mathbb{I} \otimes \Lambda)\varrho \geq 0 . \tag{5.27}$$

If the map is not CP, then this condition is not trivial, i.e. for some states $(\mathbb{I} \otimes \Lambda)\varrho$ is not positive. Consider the map given by $\Lambda(A) = (\operatorname{Tr} A)\mathbb{I} - A$. The eigenvalues of the resulting operator $\Lambda(A)$ are given by $\lambda_i = \operatorname{Tr} A - a_i$, where a_i are the eigenvalues of A. If A is positive, then $a_i \geq 0$. Now, since $\operatorname{Tr} A = \sum_i a_i$, then λ_i are also nonnegative. Thus the map is positive. Now, the formula (5.27) and the dual formula $(\Lambda \otimes \mathbb{I})\varrho \geq 0$ applied to this particular map imply that separable states must satisfy the following inequalities:

$$\mathbb{I} \otimes \varrho_B - \varrho \geq 0 , \quad \varrho_A \otimes \mathbb{I} - \varrho \geq 0 . \tag{5.28}$$

The two conditions, taken jointly, are called the reduction criterion [281, 309]. One can check that it implies the entropic inequalities (hence it is better in "detecting" entanglement). From the reduction criterion, it follows that states ϱ of a $d \otimes d$ system with $\mathcal{F}(\varrho) > 1/d$ must be entangled (this was originally argued in [66]). Indeed, from the above inequalities, it follows that for a separable state σ and a maximally entangled state ψ_{me} one has $\langle \psi_{\mathrm{me}} | \sigma_A \otimes \mathbb{I} - \sigma | \psi_{\mathrm{me}} \rangle \geq 0$. Since the reduced density matrix $\varrho_A^{\psi_{\mathrm{me}}}$ of the state ψ_{me} is proportional to the identity, we obtain $\langle \psi_{\mathrm{me}} | \sigma_A \otimes \mathbb{I} | \psi_{\mathrm{me}} \rangle = \operatorname{Tr}(\varrho_A^{\psi_{\mathrm{me}}} \sigma_A) = 1/d$. Hence we obtain $\mathcal{F} \leq 1/d$. We conclude that the latter condition is a separability criterion.

Let us note finally [281], that for $2 \otimes 2$ and $2 \otimes 3$ systems the reduction criterion is equivalent to the PPT criterion, and hence is equivalent to separability.

Strong Separability Criteria from an Entanglement Witness Consider the unitary *flip* operator V on a $d \otimes d$ system defined by $V\psi \otimes \phi = \phi \otimes \psi$. Note that it can be written as $V = P_\mathrm{S} - P_\mathrm{A}$, where P_S and P_A are projectors onto the symmetric and antisymmetric subspaces, respectively, of the total space. Hence V is a dichotomic observable (with eigenvalues ± 1). One can check that $\operatorname{Tr} VA \otimes B = \operatorname{Tr} AB$ for any operators A, B. Then V is an entanglement witness, so that $\operatorname{Tr} \varrho V \geq 0$ is a separability criterion [263]. Now, let us find the corresponding positive map via the formula (5.23). One easily finds that it is a transposition (up to an irrelevant factor). Remarkably, in this way, given an entanglement witness, one can find the corresponding map, so as to obtain the *much stronger* criterion given by (5.27).

Werner States In [263] Werner considered states that do not change if both subsystems are subjected to the same unitary transformation:

$$\varrho = U \otimes U \varrho U^\dagger \otimes U^\dagger \quad \text{for any unitary } U . \tag{5.29}$$

He showed that such states (called Werner states) must be of the following form:

$$\varrho_W(d) = \frac{1}{d^2 - \beta d}(\mathbb{I} + \beta V), \quad -1 \leq \beta \leq 1, \tag{5.30}$$

where V is the flip operator defined above. Another form for ϱ_W is [310]

$$\varrho_W(d) = p\frac{P_A}{N_A} + (1-p)\frac{P_S}{N_S}, \quad 0 \leq p \leq 1, \tag{5.31}$$

where $N_A = (d^2 - d)/2$ and $N_S = (d^2 + d)/2$ are the dimensions of the antisymmetric and symmetric subspaces, respectively. It was shown [263] that ϱ_W is entangled if and only if Tr $V\varrho_W < 0$. Equivalent conditions are $\beta < -1/d$, $p > 0$ or ϱ is NPT. Thus ϱ_W is separable if and only if it is PPT. For $d = 2$ (the two-qubit case) the state can be written as (see [264])

$$\varrho_W(2) = p|\psi_-\rangle\langle\psi_-| + (1-p)\frac{\mathbb{I}}{4}, \quad -\frac{1}{3} \leq p \leq 1. \tag{5.32}$$

Note that any state ϱ, if subjected to a random transformation of the form $U \otimes U$ (we call such an operation $U \otimes U$ *twirling*), becomes a Werner state:

$$\int dU\, U \otimes U\, \varrho\, U^\dagger \otimes U^\dagger = \varrho_W. \tag{5.33}$$

Moreover, Tr ϱV = Tr $\varrho_W V$ (i.e. Tr ϱV is invariant under $U \otimes U$ twirling).

Isotropic State If we apply a local unitary transformation to the state (5.32), changing ψ_- into ψ_+, we can generalize its form to higher dimensions as follows [281]:

$$\varrho(p,d) = pP_+ + (1-p)\frac{\mathbb{I}}{d^2}, \quad \text{where} \quad -\frac{1}{d^2 - 1} \leq p \leq 1. \tag{5.34}$$

The state will be called "isotropic" [311] here.[13] For $p > 0$ it is interpreted as a mixture of a maximally entangled state P_+ with a completely chaotic noise represented by $\mathbb{I}/d$. It was shown that it is the only state invariant under $U \otimes U^*$ transformations.[14] If we use the singlet fraction $F = \text{Tr}\, \varrho P_+$ as a parameter, we obtain

$$\varrho(F,d) = \frac{d^2}{d^2 - 1}\left((1-F)\frac{\mathbb{I}}{d^2} + (F - \frac{1}{d^2})P_+\right), \quad 0 \leq F \leq 1. \tag{5.35}$$

The two parameters are related via $p = (d^2 F - 1)/(d^2 - 1)$. The state is entangled if and only if $F > 1/d$ or, equivalently, if it is NPT. Similarly to the case for Werner states, a state subjected to $U \otimes U^*$ twirling (random $U \otimes U^*$ operations) becomes isotropic, and the parameter $F(\varrho)$ is invariant under this operation.

---
[13] In [281] it was called a "noisy singlet".
[14] The star denotes complex conjugation.

A Two-Qubit State
Consider the following two-qubit state:

$$\varrho = p|\psi_-\rangle\langle\psi_-| + (1-p)|00\rangle\langle00| . \tag{5.36}$$

From (5.9) we obtain the result that for $p \leq 1/\sqrt{2}$, the CHSH–Bell inequalities are satisfied. A little bit stronger is the criterion involving the fully entangled fraction; we have $\mathcal{F} \leq 1/2$ for $p \leq 1/2$. The entropic inequalities, apart from the one involving S_0, are equivalent to each other for this state and give again $p \leq 1/2$. Thus they reveal entanglement for $p > 1/2$. By applying the partial transposition one can convince oneself that the state is entangled for all $p > 0$ (for $p = 0$ it is manifestly separable).

Entangled PPT State Via Nondecomposable Positive Map
Consider the following state (constructed on the basis of Størmer matrices [288]) of a $3 \otimes 3$ system:

$$\sigma_\alpha = \frac{2}{7}|\psi_+\rangle\langle\psi_+| + \frac{\alpha}{7}\sigma_+ + \frac{5-\alpha}{7}\sigma_- \quad 2 \leq \alpha \leq 5 , \tag{5.37}$$

where

$$\sigma_+ = \frac{1}{3}(|0\rangle|1\rangle\langle0|\langle1| + |1\rangle|2\rangle\langle1|\langle2| + |2\rangle|0\rangle\langle2|\langle0|) ,$$

$$\sigma_- = \frac{1}{3}(|1\rangle|0\rangle\langle1|\langle0| + |2\rangle|1\rangle\langle2|\langle1| + |0\rangle|2\rangle\langle0|\langle2|) . \tag{5.38}$$

Using the formulas (5.15), one easily finds that for $\alpha \leq 4$ the state is PPT. Consider now the following map [287]:

$$\Lambda\left(\begin{bmatrix} a_{11} & a_{12} & a_{13} \\ a_{21} & a_{22} & a_{23} \\ a_{31} & a_{32} & a_{33} \end{bmatrix}\right) = \begin{bmatrix} a_{11} & -a_{12} & -a_{13} \\ -a_{21} & a_{22} & -a_{23} \\ -a_{31} & -a_{32} & a_{33} \end{bmatrix} + \begin{bmatrix} a_{33} & 0 & 0 \\ 0 & a_{11} & 0 \\ 0 & 0 & a_{22} \end{bmatrix} . \tag{5.39}$$

This map has been shown to be positive [287]. Now, one can calculate the operator $(\mathbb{I} \otimes \Lambda)\varrho$ and find that one of its eigenvalues is negative for $\alpha > 3$ (explicitly, $\lambda_- = (3-\alpha)/2$). This implies that

- the state is entangled (for a separable state we would have $(\mathbb{I} \otimes \Lambda)\varrho \geq 0$)
- the map is nondecomposable (for a decomposable map and PPT state we also would have $(\mathbb{I} \otimes \Lambda)\varrho \geq 0$).

For $2 \leq \alpha \leq 3$ it is separable, as it can be written as a mixture of other separable states $\sigma_\alpha = 6/7\varrho_1 + (\alpha-2)/7\sigma_+ + (3-\alpha)/7\sigma_-$, where $\varrho_1 = (|\psi_+\rangle\langle\psi_+| + \sigma_+ + \sigma_-)/3$. The latter state can be written as an integral over product states:

$$\varrho_1 = \frac{1}{8}\int_0^{2\pi} |\psi(\theta)\rangle\langle\psi(\theta)| \otimes |\psi(-\theta)\rangle\langle\psi(-\theta)| \frac{d\theta}{2\pi} ,$$

where $|\psi(\theta)\rangle = 1/\sqrt{3}(|0\rangle + e^{i\theta}|1\rangle + e^{-2i\theta}|2\rangle)$ (there exists also a finite decomposition exploiting phases of roots of unity [312]).

5.2.5 Volumes of Entangled and Separable States

The question of the volume of the set of separable or entangled states in the set of all states, raised in [297], is important for several reasons. First, one might be interested in the following basic question: is the world more classical or more quantum? Second, the size of the volume would reflect a consideration that is important for the numerical analysis of entanglement, that of to what extent separable or entangled states are typical. Later it became apparent that considerations of the volume of separable states led to important results concerning the question of the relevance of entanglement in quantum computing [313].

We shall mainly consider a qualitative question: is the volume of separable (V_s), entangled (V_e) or PPT entangled (V_{pe}) states nonzero? All these problems can be solved by the same method [297]: one picks a suitable state from either of the sets and tries to show that some (perhaps small) ball round the state is still contained in the set.

For separable states one takes the ball round the maximally mixed state: one needs a number p_0 such that for any state $\tilde{\varrho}$ the state

$$\varrho = p\mathbb{I}/N + (1-p)\tilde{\varrho} \tag{5.40}$$

is separable for all $p \leq p_0$ (here N is the dimension of the total system). In [297] it was shown that, in the general case of multipartite systems of any finite dimension, such a p_0 exists. Note that in fact we have obtained a *sufficient* condition for separability: if the eigenvalues of a given state do not differ too much from the uniform spectrum of the maximally mixed state, then the state must be separable. One would like to have some concrete values of p_0 that satisfy the condition (the larger p_0 is, the stronger the condition).

Consider, for example, the $2 \otimes 2$ system. Here one can provide the largest possible p_0, as there exists a necessary and sufficient condition for separability (the PPT criterion). Consider the eigenvalues λ_i of the partial transposition of the state (5.40). They are of the form $\lambda_i = (1-p)/N + p\tilde{\lambda}_i$, where $\tilde{\lambda}_i$ are the eigenvalues of $\tilde{\varrho}^{T_B}$ (in our case $N = 4$). One easily can see (on the basis of the Schmidt decomposition) that a partial transposition of a pure state cannot have eigenvalues smaller than $-1/2$. Hence the same is true for mixed states. In conclusion, we obtain the result that if $(1-p)/N - p/2 \geq 0$ then the eigenvalues λ_i are nonnegative for arbitrary $\tilde{\varrho}$. Thus for the $2 \otimes 2$ system one can take $p_0 = 1/3$ to obtain a sufficient condition for separability. Concrete values of p_0 for the case of n-partite systems of dimension d were obtained in [270]:

$$p_0 = \frac{1}{(1+2/d)^{n-1}}. \tag{5.41}$$

These considerations proved to be crucial for the analysis of the experimental implementation of quantum algorithms in high-temperature systems via nuclear magnetic resonance (NMR) methods. This is because the generic state used in this approach is the maximally mixed state with a small admixture of some pure entangled state. In [313] the sufficient conditions of the above kind were further developed, and it was concluded that in all of the NMR quantum computing experiments performed so far, the admixture of the pure state was too small. Thus the total state used in these experiments was separable: it satisfied a condition sufficient for separability. This raised an interesting discussion as to what extent entanglement is necessary for quantum computing [314, 315] (see also [316]). Even though there is still no general answer, it was shown [314] that the Shor algorithm [9] requires entanglement.

Let us now turn back to the question of the volumes of V_e and V_{pe}. If one takes ψ_+ of a $d \otimes d$ system for simplicity, it is easy to see that a not very large admixture of any state will ensure $F > 1/d$. Thus any state belonging to the neighborhood must be entangled. Showing that the volume of PPT entangled states is nonzero is a bit more involved [297].

In conclusion, all three types of states are not atypical in the set of all states of a given system. However, it appears that the ratio of the volume of the set of PPT states V_{PPT} (and hence also separable states) to the volume of the total set of states goes down exponentially with the dimension of the system (see Fig. 5.2). This result was obtained numerically [297] and still

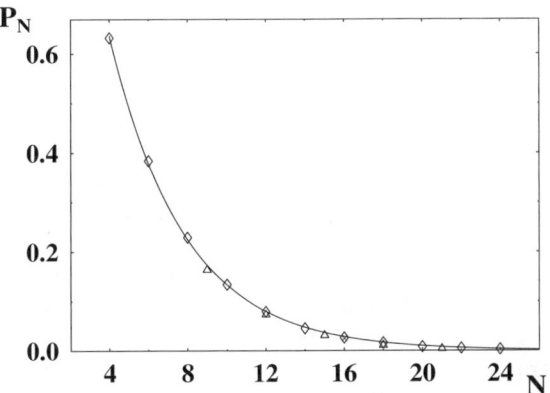

Fig. 5.2. The ratio $P_N = V_{\text{PPT}}/V$ of the volume of PPT states to the volume of the set of all states versus the dimension N of the total system. Different symbols distinguish different sizes of one subsystem ($k = 2$ ($\diamond$), $k = 3$ ($\triangle$)). (This figure is reproduced from Phys. Rev. A **58**, 883 (1998) by permission of the authors)

awaits analytical proof.[15] However, it is compatible with the rigorous result in [318] that in the infinite-dimensional case the set of separable states is nowhere dense in the total set of states. Then the generic infinite-dimensional state is entangled.

5.3 Mixed-State Entanglement as a Resource for Quantum Communication

As one knows, if two distant observers (one usually calls them Alice and Bob) share a pair of particles in a singlet state ψ_- then they can send a quantum state to one another by the use of only two additional classical bits. This is called quantum teleportation [261].[16] If the classical communication is free of charge (since it is much cheaper than communication of quantum bits), one can say that a singlet pair is a resource equivalent to sending one qubit. In the following, it will be shown that mixed-state entanglement can also be a resource for quantum communication. Quantum communication via mixed entangled states will require, apart from teleportation, an action called distillation. It will be also shown that there exists a peculiar type of entanglement (bound entanglement) that is a surprisingly weak resource.

5.3.1 Distillation of Entanglement: Counterfactual Error Correction

Now we will attempt to describe the ingenious concept of distillation of entanglement introduced in [266] and developed in [66,319] (see also [320]). To this end, let us first briefly describe the idea of classical and quantum communication via a noisy channel. As is known [321], the central idea of classical information theory, pioneered by Shannon, is that one can send information reliably and with a nonzero rate via a noisy information channel. This is achieved by coding: the k input bits of information are encoded into a larger number of n bits. Such a package is sent down the noisy channel. Then the receiver performs a decoding transformation, recovering the k input bits with asymptotically (in the limit of large n and k) perfect fidelity. Moreover, the asymptotic rate of information transmission k/n is nonzero.

In quantum domain, one would like to communicate *quantum states* instead of classical messages. It appears that an analogous scheme can be applied here [62,322]. The k input qubits of quantum information are supplemented with additional qubits in some standard initial state, and the total system of n qubits is subjected to some quantum transformation. Now the package can be sent down the channel. After the decoding operation, the

---

[15] The result could depend on the measure of the volume chosen [317]. In [298] two different measures were compared and produced similar results.

[16] It is called "entanglement-assisted teleportation" in Chap. 2 of this book.

state of k qubits is recovered with asymptotically perfect fidelity [69] (now it is *quantum* fidelity – characterizing how close the output state is to the input state), regardless of the particular form of the state. The discovery of the above possibility (called quantum error correction; we shall call it here *direct* error correction) initiated, in particular, extensive studies of quantum error-correcting codes (see [255] and references therein), as well as studies of the capacities of quantum channels (see [94] and references therein).[17] A common example of a quantum channel is the one-qubit quantum depolarizing channel: here an input state is undisturbed with probability p and subjected to a random unitary transformation with probability $1-p$. It can be described by the following completely positive map:

$$\varrho \to \Lambda(\varrho) = p\varrho + (1-p)\frac{\mathbb{I}}{2}, \qquad (5.42)$$

where $\mathbb{I}/2$ is the maximally mixed state of one qubit. This channel has been thoroughly investigated [66, 266, 323, 324]. What is important here is that it has been shown [324] that for $p \leq 2/3$ the above method of error correction does not work. In the classical domain, it would mean that the channel was useless. Here, surprisingly, there is a trick that allows one to beat this limit, even down to $p = 1/3$! The scheme that realizes this fact is quite mysterious. In direct error correction we deal directly with the systems carrying the information to be protected. Now, it appears that by using *entanglement*, one can remove the results of the action of noise without even having the information to be sent. Therefore, it can be called *counterfactual* error correction.

How does this work? The idea itself is not complicated. Instead of sending the qubits of information, Alice (the sender) sends Bob particles from entangled pairs (in the state ψ_-), keeping one particle from each pair. The pairs are disturbed by the action of the channel, so that their state turns into a mixture[18] that still possesses some residual entanglement. Now, it turns out that by local quantum operations (including collective actions over all members of the pairs in each lab) and classical communication (local operations and classical communication, LOCC) between Alice and Bob, Alice and Bob are able to obtain a smaller number of pairs in a nearly maximally entangled state ψ_- (see Fig. 5.3). Such a procedure, proposed in [266], is called distillation. As in the case of direct error correction, one can achieve a finite asymptotic rate k/n for the distilled pairs per input pair, and the fidelity, which now denotes the similarity of the distilled pairs to a product of singlet pairs, is asymptotically perfect. Now, the distilled pairs can be used for teleportation of quantum information. The maximal possible rate achievable within the above framework is called the *entanglement of distillation* of the

---

[17] The capacity Q of a quantum channel is the greatest ratio k/n for reliable transmission down the given channel.

[18] If the channel is memoryless, the mixture factorizes into states ϱ of individual pairs.

state ϱ, and is denoted by $D(\varrho)$. Thus, if Alice and Bob share n pairs each in state ϱ, they can faithfully teleport $k = nD(\varrho)$ qubits.

As we have mentioned, the error correction stage and the transmission stage are separated in time here; the error correction can be performed even before the information to be sent was produced. Using the terminology of [325], one can say that Alice and Bob operate on *potentialities* (an entangled pair represents a potential communication) and correct the potential error, so that when the *actual* information is coming, it can be teleported without any additional action.

The above scheme is not only mysterious. It is also much more powerful than the direct method. In the next section we describe a distillation protocol that allows one to send quantum information reliably via a channel with $p > 1/3$. A general question is: where are the limits of distillation? As we have seen, the basic action refers to mixed bipartite states, so that instead of talking about channels, we can concentrate on bipartite states. The question can be formulated as follows: which states ϱ can be distilled by the most general LOCC actions? Here, by saying that a *state* ϱ can be distilled, we mean that Alice and Bob can obtain singlets from the initial state $\varrho^{\otimes n}$ of n pairs (thus we shall work with memoryless channels).

One can easily see that separable states cannot be distilled: they contain no entanglement, so it is impossible to convert them into entangled states by LOCC operations. Then the final form of our question is: can all entangled states be distilled? Before the answer to this question was provided, the default was "yes", and the problem was how to prove it. Now, we know that the answer is "no", so that the structure of the entanglement of bipartite states is much more puzzling than one might have suspected.

Finally, we should mention that for pure states the problem of conversion into singlet pairs has been solved. Here there is no surprise: all entangled pure states can be distilled [326] (see also [327]). What is especially important is that this distillation can be performed reversibly: from the singlet pairs obtained, we can recover (asymptotically) the same number of input pairs [326]. As we shall see, this is not the case for mixed states.

Fig. 5.3. Distillation of mixed-state entanglement

5.3.2 Distillation of Two-Qubit States

In this section, we shall describe what was historically the first distillation protocol for two-qubit states, devised by Bennett, Brassard, Popescu, Schumacher, Smolin and Wootters (BBPSSW) [266]. Then we shall show that a more general protocol can distill any entangled two-qubit state.[19]

BBPSSW Distillation Protocol The BBPSSW distillation protocol still remains the most transparent example of distillation. It works for two-qubit states ϱ with a fully entangled fraction satisfying $\mathcal{F} > 1/2$. Such states are equivalent to those with $F > 1/2$, so that we can restrict ourselves to the latter states. Hence we assume that Alice and Bob initially share a huge number of pairs, each in the same state ϱ with $F > 1/2$, so that the total state is $\varrho^{\otimes n}$. Now they aim to obtain a smaller number of pairs with a higher singlet fraction F. To this end they iterate the following steps:

1. They take two pairs and apply $U \otimes U^*$ twirling to each of them, i.e. a random unitary transformation of the form $U \otimes U^*$ (Alice picks at random a transformation U, applies it, and communicates to Bob which transformation she has chosen; then he applies U^* to his particle). Thus one has a transformation from two copies of ϱ to two copies of the isotropic state ϱ_F with an unchanged F:

$$\varrho \otimes \varrho \to \varrho_F \otimes \varrho_F . \tag{5.43}$$

2. Each party performs the unitary transformation XOR[20] on his/her members of the pairs (see Fig. 5.4). The transformation is given by

$$U_{\text{XOR}} |a\rangle |b\rangle = |a\rangle |(a+b) \bmod 2\rangle \tag{5.44}$$

(the first qubit is called the source, the second qubit the target). They obtain some complicated state $\tilde{\varrho}$ of two pairs.

3. The pair of target qubits is measured locally in the basis $|0\rangle, |1\rangle$ and it is discarded. If the results agree (success), the source pair is kept and has a greater singlet fraction. Otherwise (failure), the source pair is discarded too.

If the results in step 3 agree, the final state ϱ' of the source pair kept can be calculated from the formula

$$\varrho' = \text{Tr}_{\mathcal{H}_t}(P_t \otimes \mathbb{I}_s \tilde{\varrho} P_t \otimes \mathbb{I}_s) , \tag{5.45}$$

---

[19] In this contribution we restrict ourselves to distillation by means of perfect operations. The more realistic case where there are imperfections in the quantum operations performed by Alice and Bob is considered in [328].

[20] The quantum XOR gate is the most common quantum two-qubit gate and was introduced in [329].

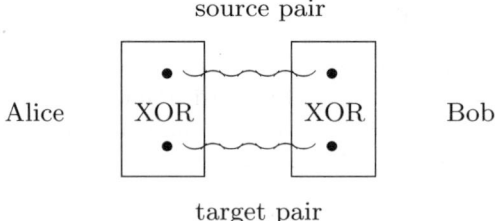

Fig. 5.4. Bilateral quantum XOR operation

where the partial trace is performed over the Hilbert space $\mathcal{H}(t)$ of the target pair, $\mathbb{I}_s$ is the identity on the space of the source pair (because it was not measured), and $P_t = |00\rangle\langle 00| + |11\rangle\langle 11|$ acts on target pair space and corresponds to the case "results agree".

Subsequently, one can calculate the singlet fraction of the surviving pair as a function of the singlet fraction of the two initial pairs, obtaining

$$F'(F) = \frac{F^2 + (1/9)(1-F)^2}{F^2 + (2/3)F(1-F) + (5/9)(1-F)^2} \ . \tag{5.46}$$

Since the function $F(F')$ is continuous, $F'(F) > F$ for $F > 1/2$ and $F'(1) = 1$, we obtain the result that by iterating the procedure, Alice and Bob can obtain a state with arbitrarily high F. Of course, the larger F is required to be, the more pairs must be sacrificed, and the less the probability p of success is. Thus if Alice and Bob start with some F_{in} and would like to end up with some higher F_{out}, the number of final pairs will be on average $k = np/2^l$, where l and p depend on F_{in} and F_{out}, and denote the number of iterations of the function $F'(F)$ required to reach F_{out} starting from F_{in}, and the probability of a string of l successful operations, respectively.

The above method allows one to obtain an arbitrarily high F, but the asymptotic rate is zero. However, if F is high enough so that $1 - S > 0$, where S is the von Neumann entropy of the state ϱ, then there exists a protocol (called hashing) that gives a nonzero rate [66]. We shall not describe this protocol here, but we note that for any state with $F > 1/2$ Alice and Bob can start by using the recurrence method to obtain $1 - S > 0$, and then apply the hashing protocol. This gives a nonzero rate for any state with $F > 1/2$. This means that quantum information can be transmitted via a depolarizing channel (5.42) only if $p > 1/3$. Indeed, one can check that if Alice send one of the particles from a pair in a state ψ_+ via the channel to Bob, then the final state shared by them will be an isotropic one with $F > 1/2$. By repeating this process, Alice and Bob can obtain many such pairs. Then distillation will allow them to use the pairs for asymptotically faithful quantum communication.

All Entangled Two-Qubit States are Distillable As was mentioned in Sect. 5.2.4, there exist entangled two-qubit states with $\mathcal{F} < 1/2$, so that no product unitary transformation can produce $F > 1/2$. Thus the BBPSSW protocol cannot be applied to all entangled two-qubit states. We shall show below that, nevertheless, all such states are distillable [303]. It was possible to solve the problem mainly because of the characterization of the entangled states as discussed in Sect. 5.2.3.

Since we are not interested in the value of the asymptotic rate, it suffices to show that by starting with pairs in an entangled state, Alice and Bob are able to obtain a fraction of them in a new state with $F > 1/2$ (and then the BBPSSW protocol will do the job). Our main tool will be the so-called filtering operation [326, 330], which involves *generalized* measurement performed by one of the parties (say, Alice) on individual pairs. This measurement consists of two outcomes $\{1, 2\}$, associated with operators W_1 and W_2 satisfying

$$W_1^\dagger W_1 + W_2^\dagger W_2 = \mathbb{I}_A \tag{5.47}$$

($\mathbb{I}_A$ and $\mathbb{I}_B$ denote identities on Alice's and Bob's systems, respectively). After such a measurement, the state becomes

$$\varrho \to \frac{1}{p_i} W_i \otimes \mathbb{I}_B \, \varrho \, W_i^\dagger \otimes \mathbb{I}_B, \quad i = 1, 2 \tag{5.48}$$

with probability $p_i = \text{Tr}(W_i \varrho W_i^\dagger)$. The condition (5.47) ensures $p_1 + p_2 = 1$. Now Alice will be interested only in one outcome (say, 1). If this outcome occurs, Alice and Bob keep the pair; otherwise they discard it (this requires communication from Alice to Bob). Then we are only interested in the operator W_1. If its norm does not exceed 1, one can always find a suitable W_2 such that the condition (5.47) is satisfied. Now, since neither the form of the final state (5.48) nor the fact whether p_1 is zero or not depends on the positive factor multiplying W_1, we are free to consider completely arbitrary filtering operators W_1. In conclusion, for any entangled state ϱ we must find an operator W such that the state resulting from the filtering $W \otimes \mathbb{I} \varrho W^\dagger \otimes \mathbb{I}/\text{Tr}(W \otimes \mathbb{I} \varrho W^\dagger \otimes \mathbb{I})$ has $F > 1/2$. Consider then an arbitrary two-qubit entangled state ϱ. From Theorem 5.2, we know that ϱ^{T_B} is not a positive operator, and hence there exists a vector ψ for which

$$\langle \psi | \varrho^{T_B} | \psi \rangle < 0 \,. \tag{5.49}$$

Now let us note that any vector ϕ of a $d \otimes d$ system can be written as $\phi = A_\phi \otimes \mathbb{I} \psi_+$, where A_ϕ is some operator. Indeed, write ϕ in a product basis: $\phi = \sum_{i,j=1}^{d} a_{ij} |i\rangle \otimes |j\rangle$. Then the matrix elements of the operator A_ϕ are given by $\langle i | A_\phi | j \rangle = \sqrt{d} a_{ij}$ (in our case, $d = 2$). Therefore the formula (5.49) can be rewritten in the form

$$\text{Tr}\big[(A_\psi^\dagger \otimes \mathbb{I} \varrho A_\psi \otimes \mathbb{I})^{T_B} P_+\big] < 0 \,. \tag{5.50}$$

Using the identity $\text{Tr } A^{T_B} B = \text{Tr } AB^{T_B}$, valid for any operators A, B, and the fact that $P_+^{T_B} = 1/dV$ (where V is the flip operator, see Sect. 5.2.4), we obtain

$$\text{Tr}\left[(A_\psi^\dagger \otimes \mathbb{I} \varrho A_\psi \otimes \mathbb{I}) V\right] < 0 . \tag{5.51}$$

We conclude that $A_\psi^\dagger \otimes \mathbb{I} \varrho A_\psi \otimes \mathbb{I}$ cannot be equal to the null operator, and hence we can consider the following state:

$$\tilde{\varrho} = \frac{A_\psi^\dagger \otimes \mathbb{I} \varrho A_\psi \otimes \mathbb{I}}{\text{Tr}(A_\psi^\dagger \otimes \mathbb{I} \varrho A_\psi \otimes \mathbb{I})} .$$

Now it is clear that the role of the filter W will be played by the operator $A_\psi^\dagger$. We shall show that $\langle \psi_- | \tilde{\varrho} | \psi_- \rangle > 1/2$, where $\psi_- = (|01\rangle - |10\rangle)/\sqrt{2}$. Then a suitable unitary transformation by Alice can convert $\tilde{\varrho}$ into a state ϱ' with $F > 1/2$.

From the inequality (5.51), we obtain

$$\text{Tr } \tilde{\varrho} V < 0 . \tag{5.52}$$

If we use the product basis $|1\rangle = |00\rangle, |2\rangle = |01\rangle, |3\rangle = |10\rangle, |4\rangle = |11\rangle$, the inequality (5.52) can be written as

$$\tilde{\varrho}_{11} + \tilde{\varrho}_{44} + \tilde{\varrho}_{23} + \tilde{\varrho}_{32} < 0 . \tag{5.53}$$

The above inequality, together with the trace condition $\text{Tr } \tilde{\varrho} = \sum_i \tilde{\varrho}_{ii} = 1$, gives

$$\langle \psi_- | \tilde{\varrho} | \psi_- \rangle = \frac{1}{2}(\tilde{\varrho}_{22} + \tilde{\varrho}_{33} - \tilde{\varrho}_{23} - \tilde{\varrho}_{32}) > \frac{1}{2} . \tag{5.54}$$

To summarize, given a large supply of pairs, each in an entangled state ϱ, Alice and Bob can distill maximally entangled pairs in the following way. First Alice applies a filtering determined by the operator $W = A_\psi^\dagger$, described above. Then Alice and Bob obtain, on average, a supply of np surviving pairs in the state $\tilde{\varrho}$ (here $p = \text{Tr } W \otimes \mathbb{I} \varrho W^\dagger \otimes \mathbb{I}$ is the probability that the outcome of Alice's measurement will be the one associated with the operator $W^\dagger$). Now Alice applies an operation $i\sigma_y$ to obtain a state with $F > 1/2$. Then they can use the BBPSSW protocol to distill maximally entangled pairs that are useful for quantum communication. Note that we have assumed that Alice and Bob know the initial state of the pairs. It can be shown that, if they do not know, they still can do the job (in the two-qubit case) by sacrificing $\sqrt{n}$ pairs to estimate the state (P. Horodecki, unpublished).

The above protocol can easily be shown to work in the $2 \otimes 3$ case. The protocol can be also fruitfully applied for the system $2 \otimes n$ if the state is NPT [331].

5.3.3 Examples

Consider the state (5.36) from Sect. 5.2.4, $\varrho = p|\psi_-\rangle\langle\psi_-| + (1-p)|00\rangle\langle00|$. It is entangled for all $p > 0$. In matrix notation we have

$$\varrho = \begin{bmatrix} 1-p & 0 & 0 & 0 \\ 0 & \frac{p}{2} & -\frac{p}{2} & 0 \\ 0 & -\frac{p}{2} & \frac{p}{2} & 0 \\ 0 & 0 & 0 & 0 \end{bmatrix}, \quad \varrho^{T_B} = \begin{bmatrix} 1-p & 0 & 0 & -\frac{p}{2} \\ 0 & \frac{p}{2} & 0 & 0 \\ 0 & 0 & \frac{p}{2} & 0 \\ -\frac{p}{2} & 0 & 0 & 0 \end{bmatrix}. \quad (5.55)$$

The negative eigenvalue of ϱ^{T_B} is $\lambda_- = 1/2\left(1 - p - \sqrt{(1-p)^2 + p^2}\right)$ and the corresponding (unnormalized) eigenvector $\psi = \lambda_-|00\rangle - (p/2)|11\rangle$, and hence we can take the filter to be of the form $W = \text{diag}[\lambda_-, -p/2]$. The new state $\tilde{\varrho}$ resulting from filtering is of the form

$$\tilde{\varrho} = \frac{1}{N}\begin{bmatrix} \lambda_-^2(1-p) & 0 & 0 & 0 \\ 0 & \frac{p^3}{8} & \frac{p^2}{4}\lambda_- & 0 \\ 0 & \frac{p^2}{4}\lambda_- & \frac{p}{2}\lambda_-^2 & 0 \\ 0 & 0 & 0 & 0 \end{bmatrix}, \quad (5.56)$$

where $N = \lambda_-^2(1-p) + p^2/8 + \lambda_-^2 p/2$. Now the overlap with ψ_-, given by $\langle\psi_-|\tilde{\varrho}|\psi_-\rangle = (p^3/8 + \lambda^2 p/2 - \lambda p^2/2)/N$, is greater than $1/2$ only if $p > 0$. The new state can be distilled by the BBPSSW protocol.

Below we shall prove that some states of higher-dimensional systems are distillable. We shall do this by showing that some LOCC operation can convert them (possibly with some probability) into an entangled two-qubit state.

Distillation of Isotropic State for $d \otimes d$ System For $F > 1/d$, an isotropic state can be distilled [281,313]. If *both* Alice and Bob apply the projector $P = |0\rangle\langle 0| + |1\rangle\langle 1|$, where $|0\rangle, |1\rangle$ are vectors from the local basis, then the isotropic state will be converted into a two-qubit isotropic state. (Note that the projectors play the role of filters; also, the filtering is successful if both Alice and Bob obtain outcomes corresponding to P.) Now, if the initial state satisfied $F > 1/d$ then the final state, as a two-qubit state, will have $F > 1/2$. Thus it is entangled and hence can be distilled.

Distillation and Reduction Criterion Any state ϱ of a $d \otimes d$ system that violates the reduction criterion (see Sect. 5.2.4) can be distilled [281]. Indeed, take a vector ψ for which $\langle\psi|\varrho_A \otimes \mathbb{I} - \varrho|\psi\rangle < 0$. It is easy to see that by applying the filter W given by $\psi = W \otimes \mathbb{I}\psi_+$, one obtains a state with $F > 1/d$. Now, the random $U \otimes U^*$ transformations will convert it into an isotropic state with the same F. As shown above, the latter state is distillable.

5.3.4 Bound Entanglement

In the light of the result for two qubits, it was naturally expected that any entangled state could be distilled. It was a great surprise when it became apparent that it was not the case. In [71] it was shown that there exist entangled states that cannot be distilled. The following theorem provides a necessary and sufficient condition for the distillability of a mixed state [71].

Theorem 5.4. *A state ϱ is distillable if and only if, for some two-dimensional projectors P, Q and for some number n, the state $P \otimes Q \varrho^{\otimes n} P \otimes Q$ is entangled.*

Remarks. (1) Note that the state $P \otimes Q \varrho^{\otimes n} P \otimes Q$ is effectively a two-qubit one as its support is contained in the $\mathbb{C}^2 \otimes \mathbb{C}^2$ subspace determined by the projectors P, Q. This means that the distillable entanglement is a *two-qubit* entanglement. (2) One can see that the theorem is compatible with the fact [326] that any *pure* state can be distilled.

As a consequence of this theorem, we obtain the following theorem [71]:

Theorem 5.5. *A PPT state cannot be distilled.*

Proof. We shall give here a proof independent of Theorem 5.4. As a matter of fact, we shall show that (i) the set of PPT states is invariant under LOCC operations [71] and (ii) it is bounded away from the maximally entangled state [311, 332]. Then, since $(\varrho^{\otimes n})^{T_B} = (\varrho^{T_B})^{\otimes n}$, we obtain the proof of the theorem. To prove (i), note that any LOCC operation can be written as [268]

$$\varrho \to \varrho' = \frac{1}{p} \sum_i A_i \otimes B_i \varrho A_i^\dagger \otimes B_i^\dagger , \qquad (5.57)$$

where p is a normalization constant interpreted as the probability of realization of the operation, and the map $\varrho \to \sum_i A_i \otimes B_i \varrho A_i^\dagger \otimes B_i^\dagger$ does not increase the trace (this ensures $p \leq 1$). Suppose now that ϱ is PPT, i.e. $\varrho^{T_B} \geq 0$, and examine partial transposition of the state ϱ'. We shall use the following property of partial transposition:

$$(A \otimes B \varrho C \otimes D)^{T_B} = A \otimes D^T \varrho^{T_B} C \otimes B^T \qquad (5.58)$$

for any operators A, B, C, D and ϱ. Then we obtain

$$(\varrho')^{T_B} = \sum_i A_i \otimes (B_i^\dagger)^T \varrho^{T_B} A_i \otimes (B_i)^T . \qquad (5.59)$$

Thus $(\varrho')^{T_B}$ is a result of the action of some completely positive map on an operator ϱ^{T_B} that by assumption is positive. Then also the operator $(\varrho')^{T_B}$ must be positive. Thus a LOCC map does not move outside the set of PPT states.

To prove (ii), let us now show that PPT states can never have a high singlet fraction F. Consider a PPT state ϱ of a $d \otimes d$ system. We obtain

$$\text{Tr } \varrho P_+ = \text{Tr } \varrho^{T_B} P_+^{T_B} . \tag{5.60}$$

Now, it is easy to check that $P_+ = 1/dV$, where V is the flip operator described in Sect. 5.2.4. Note that V is Hermitian and has eigenvalues ± 1. Since ϱ is PPT then $\tilde{\varrho} = \varrho^{T_B}$ is a legitimate state, and the above expression can be rewritten in terms of the mean value of the observable V as

$$\text{Tr } \varrho P_+ = \frac{1}{d} \text{Tr } \tilde{\varrho} V . \tag{5.61}$$

The mean value of a dichotomic observable cannot exceed 1, so that we obtain

$$F(\varrho) \leq \frac{1}{d} . \tag{5.62}$$

Thus the maximal possible singlet fraction that can be attained by PPT states is the one that can be obtained without any prior entanglement between the parties. Indeed, a *product* state $|00\rangle$ has a singlet fraction $1/d$ (if it belongs to the Hilbert space $C^d \otimes C^d$). Consequently, for however large an amount of PPT pairs, even a single two-qubit pair with $F > 1/2$ cannot be obtained by LOCC actions. □

Now, one can appreciate the results presented in the first part of this contribution. From Sect. 5.2.3, we know that there exist entangled states that are PPT. So far, the question of whether there exist entangled states that are PPT has been merely a technical one. At this point, since the above theorem implies that PPT states are nondistillable, we can draw a remarkable conclusion: there exist nondistillable entangled states. Since, in the process of distillation, no entanglement can be liberated to the useful singlet form, they have been called *bound entangled (BE) states*. Thus there exist at least two *qualitatively different* types of entanglement: apart from the *free* entanglement that can be distilled, there is a *bound* one that cannot be distilled and seems to be completely useless for quantum communication. This *discontinuity* of the structure of the entanglement of mixed states was considered to be possible for multipartite systems, but it was completely surprising for bipartite systems. It should be emphasized here that the BE states are not atypical in the set of all possible states: as we have mentioned in Sect. 5.2.5, the volume of the PPT entangled states is nonzero. One of the main consequences of the existence of BE states is that it reveals a transparent form of *irreversibility* in entanglement processing. If Alice and Bob share pairs in a pure state, then to produce a BE state they need some prior entanglement.[21] However, once they have produced the BE states, they are not able to recover the pure entanglement from them. It is *entirely* lost. This is a *qualitative* irreversibility, which is probably a source of the *quantitative* irreversibility [66,

---

[21] This was rigorously proved recently in [333].

266] that is due to the fact that we need more pure entanglement to produce some mixed states than we can then distill back from them [269, 334].[22]

To analyse the phenomenon of bound entanglement, one needs as many examples of BE states as possible. Hence there is a very exciting physical motivation for the search for PPT entangled states. In Sect. 5.2.3 we discussed different methods of searching . As a result we obtained a couple of examples of BE states via the separability criterion given by Theorem 5.3, from the mathematical literature on nondecomposable maps and via unextendible product bases.

The examples produced via UPBs are extremely interesting from the physical point of view. This is because a UPB is not only a mathematical object: as shown in [273], it produces a very curious physical effect [120] called "nonlocality without entanglement". Namely, suppose that Alice and Bob share a pair in one of the states from the UPB, but they do not know which state this is. It appears that by LOCC operations (with finite resources), they are not able to read the identity of the state. However, if the particles were together, then, since the states are orthogonal, they could be perfectly distinguished from each other. Thus we have a highly *nonclassical* effect produced by an ensemble of *separable* states. On the other hand, the BE state associated with the given UPB (the uniform state on the complementary subspace, see (5.26)) presents opposite features: it *is* entangled but, since its entanglement is bound, it ceases to behave in a quantum manner. Moreover, in both situations we have a kind of irreversibility. As was mentioned, BE states are a reflection of the formation–distillation irreversibility: to create them by LOCC from singlet pairs, Alice and Bob need a nonzero amount of the latter. However, once they are created, there is no way to distill singlets out of them. On the other hand, a UPB exhibits preparation–measurement irreversibility: any of the states belonging to the UPB can be prepared by LOCC operations, but once Alice and Bob forget the identity of the state, they cannot recover it by LOCC. This surprising connection between some BE states and bases that are not distinguishable by LOCC implies many interesting questions concerning the future unification of our knowledge about the nature of quantum information.

Finally, we shall mention a result concerning the *rank* of the BE state. In numerical analysis of BE states (especially their tensor products), it is very convenient to have examples with low rank. However, in [282] the following bound on the rank of the BE state ϱ was derived:

$$R(\varrho) \geq \max\{R(\varrho_A), R(\varrho_B)\} \ . \tag{5.63}$$

---

[22] In fact, the existence of both kinds of irreversibility has not been rigorously proved so far (see [335]). The proof of quantitative irreversibility in [336] turned out to be invalid: it was based on a theorem [311] on the additivity of the relative entropy of Werner states. However, an explicit counterexample to this theorem was provided in [337].

(Recall that $R(\varrho)$ denotes the rank of ϱ.) Note that the above inequality is nothing but the entropic inequality (5.10) with the entropy (5.11). Thus it appears that the latter inequality is a necessary condition not only for separability, but also for nondistillability. The proof is based on the fact [281] that any state violating a reduction criterion (see Sects. 5.2.4 and 5.3.3) can be distilled. It can be shown that, if a state violates the above equation, then it must also violate a reduction criterion, and hence can be distilled. Then it follows that there does not exist any BE state of rank two [282]. Indeed, if such a state existed, then its local ranks could not exceed two. Hence the total state would be effectively a two-qubit state. However, from Sect. 5.3.2 we know that two-qubit bound entangled states do not exist.

5.3.5 Do There Exist Bound Entangled NPT States?

So far we have considered BE states that arise from Theorem 5.5, which says that the NPT condition is necessary for distillability. As mentioned in Sect. 5.3.2, for $2 \otimes n$ systems all NPT states can be distilled [331], and hence the condition is also sufficient in this case. However, it is not known whether it is sufficient in general. A necessary and sufficient condition is given by Theorem 5.4. To find if this condition is equivalent to the PPT one, it must be determined whether there exists an NPT state such that, for any number of copies n, the state $\varrho^{\otimes n}$ will not have an entangled two-qubit "substate" (i.e. the state $P \otimes Q \varrho^{\otimes n} P \otimes Q$). In [281] it was pointed out that one can reduce the problem by means of the following observation.

Proposition 5.1. *The following statements are equivalent:*

1. *Any NPT state is distillable.*
2. *Any entangled Werner state (5.30) is distillable.*

Proof. The proof of the implication (1) $\Rightarrow$ (2) is immediate, as Werner states are entangled if and only if they are NPT. If we can distill any NPT state, then also Werner entangled states are distillable. To obtain (2) $\Rightarrow$ (1), note that the reasoning of Sect. 5.3.2, from (5.49) to (5.52), is insensitive to the dimension d of the problem. Consequently, from any NPT state, a suitable filtering produces a state $\tilde{\varrho}$ satisfying Tr $\tilde{\varrho} V < 0$. As mentioned in Sect. 5.2.4, the parameter Tr ϱV is invariant under $U \otimes U$ twirling, so that by applying the latter (which is an LOCC operation), Alice and Bob obtain a Werner state ϱ_W satisfying Tr $\varrho_W V < 0$. Thus any NPT state can be converted by means of LOCC operations into an entangled Werner state, which completes the proof. □

The above proposition implies that to determine whether there exist NPT bound entangled states, one can restrict oneself to the family of Werner states, which is a one-parameter family of very high symmetry. Even after such a reduction of the problem, the latter remains extremely difficult. In [310, 338] the authors examine the nth tensor power of Werner states (in [338] a larger, two-parameter family is considered). The results, though not conclusive yet, strongly suggest that there exist NPT bound entangled states (see Fig. 5.5).

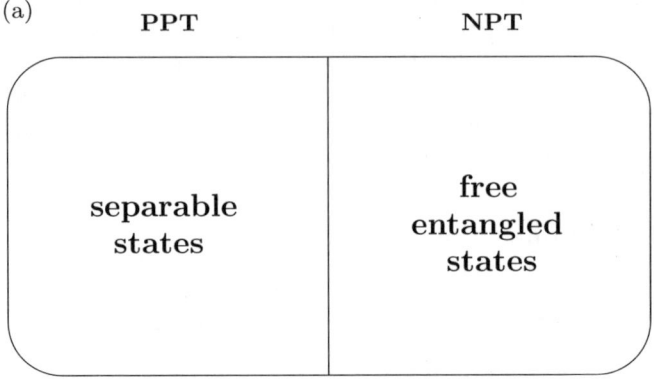

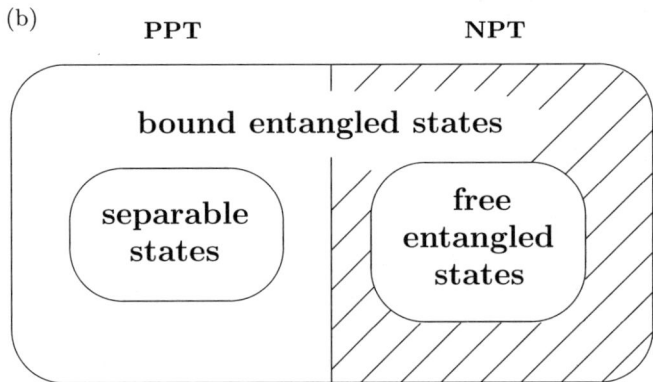

Fig. 5.5. Entanglement and distillability of mixed states for $2 \otimes 2$ and $2 \otimes 3$ systems (**a**) and for higher dimensions (**b**). The *area filled with diagonal lines* denotes the hypothetical set of bound entangled NPT states

Thus it is likely that the characterization of distillable states is not as simple as reduction to a NPT condition. The possible existence of NPT bound entanglement would make the total picture much more obscure (and hence much more interesting). Among others, there would arise the following question: for two distinct BE states ϱ_1 and ϱ_2, is the state $\varrho_1 \otimes \varrho_2$ also BE? (If BE was equivalent to PPT, this question would have an immediate answer "yes", because the PPT property is additive, i.e. if two states are PPT, then so is their tensor product [284]). Recently, a negative answer to this question was obtained in [339] in the case of a multipartite system. For bipartite states the answer is still unknown.

5.3.6 Example

Consider the family of states (5.37) considered in Sect. 5.2.4. One obtains the following classification: ϱ is

- separable for $2 \leq \alpha \leq 3$
- bound entangled (BE) for $3 < \alpha \leq 4$
- free entangled (FE) for $4 < \alpha \leq 5$.

The separability was shown in Sect. (5.2.4). It was also shown there that, for $3 < \alpha \leq 4$, the state is entangled and PPT. In this case we conclude that it is BE. For $\alpha > 4$, Alice and Bob can apply local projectors $P = |0\rangle\langle 0| + |1\rangle\langle 1|$, obtaining an entangled two-qubit state. Hence the initial state is FE in this region of α.

5.3.7 Some Consequences of the Existence of Bound Entanglement

A basic question that arises in the context of bound entanglement is: what is its role in quantum information theory? We shall show in the following sections that even though it is indeed a very poor type of entanglement, it can produce a nonclassical effect, enhancing quantum communication via a subtle activation-like process [340]. This will lead us to a new paradigm of entanglement processing that extends the "LOCC paradigm". Moreover, the existence of bound entanglement means that there exist stronger limits on the distillation rate than were expected before. We shall report these and other consequences in the next few subsections.

Bound Entanglement and Teleportation By definition, BE states cannot be distilled, and hence it is impossible to obtain faithful teleportation via such states. However, it might be the case that the transmission fidelity of *imperfect* teleportation might still be better than that achievable with a purely classical channel, i.e. without sharing any entanglement (this is a way of revealing a manifestation of quantum features of some mixed states [264]).[23] Initial searches produced a negative result [314]. Here we present more general results, according to which the most general teleportation scheme cannot produce better than classical fidelity if Alice and Bob share BE states.

---

[23] For a detailed study of the standard teleportation scheme via a mixed two-qubit state, see [341]. The optimal one-way teleportation scheme via pure states was obtained in [342].

General Teleportation Scheme Teleportation, as originally devised [261], is a way of transmitting a quantum state by use of a classical channel and a bipartite entangled state (pure singlet state) shared by Alice and Bob. The most general scheme of teleportation would then be of the following form [343]. There are three systems: that of the input particle, the state of which is to be teleported (we ascribe to this system the Hilbert space $\mathcal{H}_{A'}$), and two systems that are in the entangled state ϱ_{AB} (with Hilbert space $\mathcal{H} = \mathcal{H}_A \otimes \mathcal{H}_B$). For simplicity we assume that $\dim \mathcal{H}_{A'} = \dim \mathcal{H}_A = \dim \mathcal{H}_B = d$. The initial state is

$$|\psi_{A'}\rangle\langle\psi_{A'}| \otimes \varrho_{AB},$$

where $\psi_{A'}$ is the state to be teleported (unknown to Alice and Bob). Now Alice and Bob perform some trace-preserving LOCC operation (trace-preserving, because teleportation is an operation that must be performed with probability 1). The form of the operation depends on the state ϱ_{AB} that is known to Alice and Bob, but is independent of the input state $\psi_{A'}$ because that state is unknown. Now the total system is in a new, perhaps very complicated state $\varrho_{A'AB}$. The transmitted state is given by $\text{Tr}_{A'A}(\varrho_{A'AB})$. The overall transmission stages are the following:

$$\psi_{A'} \to |\psi_{A'}\rangle\langle\psi_{A'}| \otimes \varrho_{AB} \to \Lambda(|\psi_{A'}\rangle\langle\psi_{A'}| \otimes \varrho_{AB}) \to \text{Tr}_{A'A}\varrho_{A'AB} = \varrho_B .$$

The *transmission fidelity* is now defined by

$$f = \overline{\langle\psi_{A'}|\varrho_B|\psi_{A'}\rangle},$$

where the average is taken over a uniform distribution of the input states $\psi_{A'}$.[24] In the original teleportation scheme (where ϱ_{AB} is a maximally entangled state), the state ϱ_B is exactly equal to the input state, so that $f = 1$. If Alice and Bob share a pair in a separable state (or, equivalently, share *no* pair), then the best one can do is the following: Alice measures the state and sends the results to Bob [264]. Since it is impossible to find the form of the state when one has only a single system in that state [345] (it would contradict the no-cloning theorem [346] (see Sect. 1)), the performance of such a process will be very poor. One can check that the best possible fidelity is $f = 2/(d+1)$. If the shared pair is entangled but is not a pure maximally entangled state, we shall obtain some intermediate value of f.

Optimal Teleportation Having defined the general teleportation scheme, one can ask about the maximal fidelity that can be achieved for a given state

---

[24] Note that the fidelity so defined is not a unique criterion of the performance of teleportation. For example, one can consider a restricted input: Alice receives one of *two* nonorthogonal vectors with some probabilities [344]. Then the formula for the fidelity would be different. In general, the fidelity is determined by a chosen distribution over input states.

ϱ_{AB} within the scheme. Thus, for a given ϱ_{AB} we must maximize f over all possible trace-preserving LOCC operations. The problem is, in general, extremely difficult. However, the high symmetry of the chosen fidelity function allows one to reduce it in the following way. It has been shown [343] that the best Alice and Bob can do is the following. They first perform some LOCC action that aims at increasing $F(\varrho_{AB})$ as much as possible. Then they perform the standard teleportation scheme, via the new state ϱ'_{AB} (just as if it were the state P_+). The fidelity obtained is given by

$$f_{\max} = \frac{F_{\max}d + 1}{d + 1}, \qquad (5.64)$$

where $F_{\max} = F(\varrho'_{AB})$ is the maximal F that can be obtained by trace-preserving LOCC actions if the initial state is ϱ_{AB}.

Teleportation Via Bound Entangled States According to (5.64), to check the performance of teleportation via BE states of a $d \otimes d$ system, we should find the maximal F attainable from BE states via trace-preserving LOCC actions. As was argued in Sect. 5.3.4, a BE state subjected to any LOCC operation remains BE. Moreover, the singlet fraction F of a BE state of a $d \otimes d$ system satisfies $F \leq 1/d$ (because states with $F > 1/d$ are distillable, as shown in Sect. 5.3.3). We conclude that, if the initial state is BE, then the highest F achievable by any (not only trace-preserving) LOCC actions is $F = 1/d$. However, as we have argued, this gives a fidelity $f = 2/(d+1)$, which can be achieved without entanglement. Thus the BE states behave here like separable states – their entanglement does not manifest itself.

Activation of Bound Entanglement Here we shall show that bound entanglement can produce a nonclassical effect, even though the effect is a very subtle one. This effect is the so-called *activation of bound entanglement* [340]. The underlying concept originates from a formal entanglement–energy analogy developed in [71, 269, 325, 335, 347]. One can imagine that the bound entanglement is like the energy of a system confined in a shallow potential well. Then, as in the process of chemical activation, if we add a small amount of extra energy to the system, its energy can be liberated.

In our case, the role of the system is played by a huge amount of bound entangled pairs, while that of the extra energy is played by a *single* pair that is free entangled. More specifically, we shall show that a process called *conclusive teleportation* [348] can be performed with arbitrarily high fidelity if Alice and Bob can perform joint operations over the BE pairs and the FE pair. We shall argue that it is impossible if either of the two elements is lacking.

Conclusive Teleportation Suppose that Alice and Bob have a pair in a state for which the optimal teleportation fidelity is f_0. Suppose, further, that the fidelity is too poor for some of Alice and Bob's purposes. What they can do to change the situation is to perform a so-called conclusive teleportation. Namely, they can perform some LOCC operation with two final outcomes 0 and 1. If they obtain the outcome 0, they fail and decide to discard the pair. If the outcome is 1 they perform teleportation, and the fidelity is now better than the initial f_0. Of course, the price they must pay is that the probability of success (outcome 1) may be small. The scheme is illustrated in Fig. 5.6.

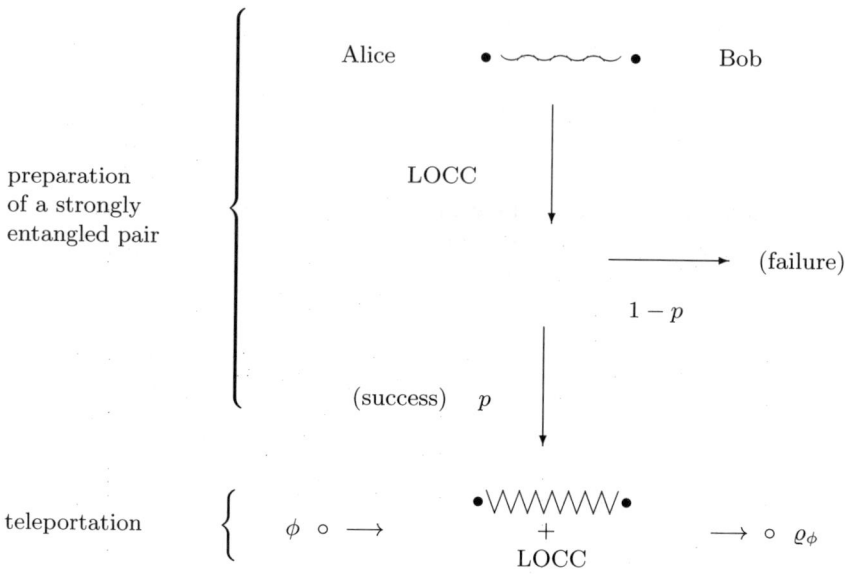

Fig. 5.6. Conclusive teleportation. Starting with a weakly entangled pair, Alice and Bob prepare with probability p a strongly entangled pair and then perform teleportation

A simple example is the following. Suppose that Alice and Bob share a pair in a pure state $\psi = a|00\rangle + b|11\rangle$ which is nearly a product state (e.g. a is close to 1). Then the standard teleportation scheme provides a rather poor fidelity $f = 2(1 + ab)/3$ [341, 349]. However, Alice can subject her particle to a filtering procedure [326, 330] described by the operation

$$\Lambda = W(\cdot)W^\dagger + V(\cdot)V^\dagger \,, \tag{5.65}$$

where $W = \text{diag}(b, a)$, $V = \text{diag}(a, b)$. Here the outcome 1 (success) corresponds to the operator W. Indeed, if this outcome is obtained, the state collapses to the singlet state

$$\tilde{\psi} = \frac{W \otimes \mathbb{I}\psi}{||W \otimes \mathbb{I}\psi||} = \frac{1}{\sqrt{2}}(|00\rangle + |11\rangle) \,. \tag{5.66}$$

Then, in this case, perfect teleportation can be performed. Thus, if Alice and Bob teleported directly via the initial state, they would obtain a very poor performance. Now they have a small but nonzero chance of performing perfect teleportation.

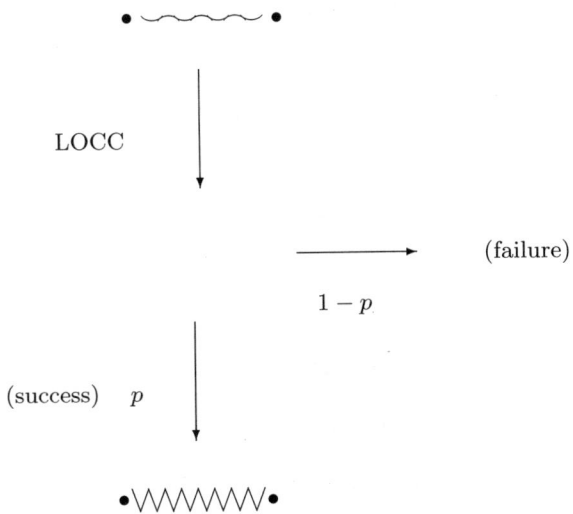

Fig. 5.7. Conclusive increase of the singlet fraction. Alice and Bob obtain, with a probability p of success, a state with a higher singlet fraction than that of the initial state

Similarly to the usual form of teleportation, conclusive teleportation can be reduced to conclusively increasing F (illustrated in Fig. 5.7), followed by the original teleportation protocol. If in the first stage Alice and Bob obtain a state with some F, then the second stage will produce the corresponding fidelity $f = (Fd + 1)/(d + 1)$. Thus we can restrict our consideration to conclusively increasing the singlet fraction. The latter process was developed in [343, 350]. An interesting peculiarity of conclusively increasing the singlet fraction is that sometimes it is impossible to obtain $F = 1$, but it is still possible to make F arbitrarily close to 1. However, if $F \to 1$, then the probability of success tends to 0, so that, indeed, it is impossible to reach $F = 1$ [343].

Activation Protocol Suppose that Alice and Bob share a *single* pair of spin-1 particles in the following free entangled mixed state:

$$\varrho_{\text{free}} = \varrho(F) \equiv F|\psi_+\rangle\langle\psi_+| + (1 - F)\sigma_+, \quad 0 < F < 1, \tag{5.67}$$

where $\sigma_\pm$ are separable states given by (5.38). It is easy to see that the state (5.67) is free entangled. Namely, after action of the local projections

$(|0\rangle\langle 0| + |1\rangle\langle 1|) \otimes (|0\rangle\langle 0| + |1\rangle\langle 1|)$, we obtain an entangled $2 \otimes 2$ state (its entanglement can be revealed by calculating partial transposition). Thus, according to Theorem 5.4, the state (5.67) is FE. By complicated considerations one can show [343] that there is a threshold $F_0 < 1$ that cannot be exceeded in the process of conclusively increasing the singlet fraction. In other words, Alice and Bob have no chance of obtaining a state ϱ' with $F(\varrho') > F_0$ (we do not know the value F_0, we only know that such a number exists).

Suppose now that Alice and Bob share, in addition, a very large number of pairs in the following BE state (the one considered in Sect. (5.3.6)):

$$\sigma_\alpha = \frac{2}{7}|\psi_+\rangle\langle\psi_+| + \frac{\alpha}{7}\sigma_+ + \frac{5-\alpha}{7}\sigma_- . \tag{5.68}$$

As stated in Sect. 5.3.6, for $3 < \alpha \leq 4$ the state is BE. As we know, there is no chance of obtaining even a pair with $F > 1/3$ from BE pairs of a $3 \otimes 3$ system. Now, it turns out that, if Alice and Bob have *both* an FE pair and the BE pairs, they can apply a simple protocol to obtain an F *arbitrarily close* to 1. Thus, owing to the connection between conclusive increasing of the singlet fraction and conclusive teleportation, the fidelity of the latter can be arbitrarily close to unity only if both an FE pair and BE pairs are shared.

The protocol [340] is similar to the recurrence distillation protocol described in Sect. 5.3.2. It is an iteration of the following two steps:

(i) Alice and Bob take the free entangled pair, in the state $\varrho_{\text{free}}(F)$, and one of the pairs, which is in the state σ_α. They perform the bilateral XOR operation $U_{\text{BXOR}} \equiv U_{\text{XOR}} \otimes U_{\text{XOR}}$, each of them treating the member of the free entangled pair as a source and the member of the bound entangled pair as a target.[25]

(ii) Alice and Bob measure the members of the source pair in the basis $|0\rangle, |1\rangle, |2\rangle$. Then they compare their results via classical communication. If the compared results differ from one another, they have to discard both pairs, and then the trial of the improvement of F fails. If the results agree, then the trial succeeds and they discard only the target pair, coming back with (as we shall see) an improved source pair to the first step (i).

After some algebra, one can see that the success in the step (ii) occurs with a nonzero probability

$$P_{F \to F'} = \frac{2F + (1-F)(5-\alpha)}{7} , \tag{5.70}$$

---

[25] Here we need the quantum XOR gate not for two qubits, as in Sect. 5.3.2, but for two qutrits (three-level systems). A general XOR operation for a $d \otimes d$ system, which was used in [281,351], is defined as

$$U_{\text{XOR}}|a\rangle|b\rangle = |a\rangle|(b+a)\bmod d\rangle , \tag{5.69}$$

where the initial states $|a\rangle$ and $|b\rangle$ correspond to the source and target states, respectively.

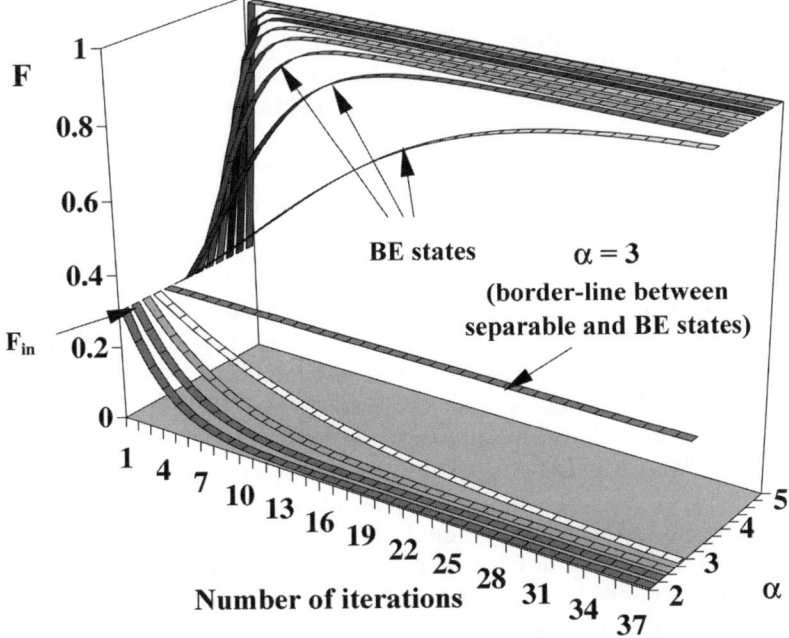

Fig. 5.8. Liberation of bound entanglement. The singlet fraction of the FE state is plotted versus the number of successful iterations of (i) and (ii), and the parameter α of the state ϱ_α of the BE pairs used. The initial singlet fraction of the FE pair is taken as $F_{\text{in}} = 0.3$ (This figure is reproduced from Phys. Rev. Lett. **82**, 1056 (1999) by permission of the authors)

leading to the transformation $\varrho(F) \to \varrho(F')$, where the improved fidelity is

$$F'(F) = \frac{2F}{2F + (1-F)(5-\alpha)}. \tag{5.71}$$

If $\alpha > 3$, then the above continuous function of F exceeds the value of F on the whole region $(0,1)$. Thus the successful repetition of the steps (i) and (ii) produces a sequence of source fidelities $F_n \to 1$. In Fig. 5.8 we have plotted the value of F obtained versus the number of iterations of the protocol and the parameter α. For $\alpha \leq 3$ the singlet fraction goes down: separable states cannot help to increase it. We can see the dramatic qualitative change at the "critical"[26] point that occurs at the borderline between separable states and bound entangled ones ($\alpha = 3$). On the other hand, it is surprising that there

---

[26] The term "critical" that we have used here reflects the rapid character of the change (see [307] for a similar "phase transition" between separable and FE states). On the other hand, the present development of thermodynamic analogies in entanglement processing [71, 268, 325, 335, 347] allows us to hope that in future one will be able to build a synthetic theory of entanglement based on thermodynamic analogies: then the "critical" point would become truly critical.

is no qualitative difference between the behavior of BE states ($3 < \alpha \leq 4$) and FE states ($4 < \alpha \leq 5$). Here the change is only quantitative, while the shape of the corresponding curves is basically the same. To our knowledge, this is the only effect we know about where bound entanglement manifests its quantumness.[27] Since the effect is very subtle, one must conclude that bound entanglement is essentially different from free entanglement and is enormously weak. For recent results on the activation effect in the multiparticle cases see [353].

Entanglement-Enhanced LOCC Operations The activation effect suggests that we should extend the paradigm of LOCC operations by including quantum communication (under suitable control). Then we obtain *entanglement-enhanced* LOCC (LOCC + EE) operations (see [354]). For example, if we allow LOCC operations and an arbitrary amount of shared bound entanglement, we obtain the LOCC + BE paradigm. One can now ask about the entanglement of formation[28] and distillation in this regime. Since BE states contain entanglement, even though it is very weak, then an infinite amount of bound entanglement could make $D_{\rm LOCC+BE}$ much larger than the usual $D_{\rm LOCC}$: one might expect $D_{\rm LOCC+BE}$ to be the maximal possible, independently of the input state ϱ [355] (e.g. for two-qubit pairs, we would have $D_{\rm LOCC+BE} = 1$ for any state). In [305, 356] it was shown that this is impossible. The argument of [305] is as follows. First, the authors recall that $D_{\rm LOCC} \leq E^F_{\rm LOCC}$ [66]. Otherwise, it would be possible to increase entanglement by means of LOCC actions. Indeed, suppose that for some state ϱ we have $D_{\rm LOCC}(\varrho) > E^F_{\rm LOCC}(\varrho)$. Then Alice and Bob could take n two-qubit pairs in a singlet state and produce $n/E^F_{\rm LOCC}$ pairs of the state ϱ. Then they could distill $n(D_{\rm LOCC}/E^F_{\rm LOCC})$ singlets, which would be greater number than n. A similar argument is applied to LOCC+ BE actions: the authors show that it is impossible to increase the number of singlet pairs by LOCC + BE actions, and conclude that $D_{\rm LOCC+BE} \leq E^F_{\rm LOCC+BE}$. On the other hand, obviously we have $E^F_{\rm LOCC+BE} \leq E^F_{\rm LOCC}$. Combining the inequalities, we obtain the result that $D_{\rm LOCC+BE}$ is bounded by the usual entanglement of formation $E^F_{\rm LOCC}$, which is maximal only for singlet-type states. A different argument in [356] is based on the results of Rains [311] on a bound for the distillation of entanglement (see Sect. 5.3.7). Thus, even if an infinite amount of BE pairs is employed, LOCC + BE operations are not enormously powerful. However, it is still possible that they are better than LOCC operations themselves, i.e. we can conjecture that $D_{\rm LOCC+BE}(\varrho) > D_{\rm LOCC}(\varrho)$ for some states ϱ.

---

[27] In the multipartite case, two other effects have recently been found [339, 352].
[28] The entanglement of formation $E^F_{\rm LOCC}(\varrho)$ of a state ϱ is the number of input singlet pairs per output pair needed to produce the state ϱ by LOCC operations [66].

Bounds for Entanglement of Distillation Bound entanglement is an achievement in qualitative description; however, as we could see in the previous section, it also has an impact on the quantitative approach. Here we shall see that it has helped to obtain a strong upper bound for the entanglement of distillation D (recall that the latter has the meaning of the capacity of a noisy teleportation channel constituted by bipartite mixed states, and hence is a central parameter of quantum communication theory).

The first upper bound for D was the entanglement of formation [66], calculated explicitly for two-qubit states [301]. However, a stronger bound has been provided in [269] (see also [334]). It is given by the following measure of entanglement [268, 269] based on the relative entropy:

$$E_{\rm VP}(\varrho) = \inf_{\sigma} S(\varrho|\sigma) , \qquad (5.72)$$

where the infimum is taken over all separable states σ. The relative entropy is defined by

$$S(\varrho|\sigma) = {\rm Tr}\, \varrho \log \varrho - {\rm Tr}\, \varrho \log \sigma .$$

Vedral and Plenio provided a complicated argument [269] showing that $E_{\rm VP}$ is an upper bound for $D(\varrho)$, under the additional assumption that it is additive. Even though we still do not know if it is indeed additive, Rains showed [311] that it is a bound for D even without this assumption. He also obtained a stronger bound by use of BE states (more precisely, PPT states). It appears that, if the infimum in (5.72) is taken over PPT states (which are bound entangled), the new measure $E_{\rm R}$ is a bound for the distillable entanglement, too. However, since the set of PPT states is strictly greater than the set of separable states, the bound is stronger. For example, the entangled PPT states have zero distillable entanglement. Since they are not separable, $E_{\rm VP}$ does not vanish for them, and hence the evaluation of D by means of $E_{\rm VP}$ is too rough. The Rains measure vanishes for these states.

We will not provide here the original proof of the Rains result. Instead we demonstrate a general theorem on bounds for distillable entanglement obtained in [357], which allows a major simplification of the proof of the result.

Theorem 5.6. *Any function B satisfying the conditions (a)–(c) below is an upper bound for the entanglement of distillation:*
(a) Weak monotonicity: $B(\varrho) \geq B[\Lambda(\varrho)]$, where Λ is a superoperator realizable by means of LOCC operations.
(b) Partial subadditivity: $B(\varrho^{\otimes n}) \leq nB(\varrho)$.
(c) Continuity for an isotropic state $\varrho(F, d)$: suppose that we have a sequence of isotropic states $\varrho(F_d, d)$ (see Sect. 5.2.4, (5.34)) such that $F_d \to 1$ if $d \to \infty$. Then we require

$$\lim_{d \to \infty} \frac{1}{\log d} B[\varrho(F_d, d)] \to 1 . \qquad (5.73)$$

Remark. If, instead of LOCC operations, we take another class C of operations including classical communication in at least one direction (e.g. the LOCC + BE operations mentioned previously), the proof, mutatis mutandis, also applies. (The condition (a) then involves the class C)

Proof. The main idea of the proof is to exploit the monotonicity condition. We shall show that if D were greater than B then, during the distillation protocol, the function B would have to increase. But this cannot be so, because distillation is a LOCC action, and hence B would violate the assumption (a). By subadditivity, we have

$$B(\varrho) \geq \frac{1}{n} B(\varrho^{\otimes n}) . \tag{5.74}$$

Distillation of n pairs aims at obtaining k pairs, each in nearly a singlet state. The asymptotic rate is $\lim k/n$. It was shown [311] that one can equally well think of the final $d \otimes d$ system as being in a state close to P_+^d. The asymptotic rate is now $\lim(\log d)/n$. Then the only relevant parameters of the final state ϱ_{out} are the dimension d and the fidelity $F(\varrho_{\text{out}})$. Thus the distillation protocol can be followed by $U \otimes U^*$ twirling, producing an isotropic final state $\varrho(d, F)$ (see Sect. 5.2.4). By condition (a), distillation does not increase B, and hence

$$\frac{1}{n} B(\varrho^{\otimes n}) \geq \frac{1}{n} B[\varrho(F_{d_n}, d_n)] . \tag{5.75}$$

Now, in the distillation process $F \to 1$, and if we consider an optimal protocol, then $(\log d)/n \to D$. Hence, by condition (c), the right-hand side of the inequality tends to $D(\varrho)$. Thus we obtain the result that $B(\varrho) \geq D(\varrho)$. □

We should check, whether the Vedral–Plenio and Rains measures satisfy the assumptions of the theorem. Subadditivity and weak monotonicity are immediate consequence of the properties of the relative entropy used in the definition of E_R (subadditivity was proved in [268], and weak monotonicity in [269]). The calculation of E_R for an isotropic state is a little bit more involved, but by using the high symmetry of the state, it was found to be [311] $E_{\text{VP}}[\varrho(F, d)] = E_R[\varrho(F, d)] = \log d + F \log F + (1-F) \log[(1-F)/(d-1)]$. Evaluating now this expression for large d, we easily find that the condition (c) is satisfied. The argument applies without any change to the Rains bound.

Finally, let us note that the Rains entanglement measure attributes no entanglement to some entangled states (the PPT entangled ones). Normally we would require that a natural postulate for an entanglement measure would be that the entanglement measure should vanish if and only if the state is separable. However, then we would have to remove distillable entanglement from the set of measures. Indeed, the distillable entanglement vanishes for some manifestly entangled states – bound entangled ones. Now the problem is: should we keep the postulate, or keep D as a good measure?

It is reasonable to keep D as a good measure, as it has a direct physical sense: it describes entanglement as a resource for quantum communication. If it is not a measure, then we must conclude that we are not interested in measures. Consequently, we adopt as a main "postulate" for an entanglement measure the following statement: "Distillable entanglement is a good measure". So we must abandon the postulate. The apparent paradox can be removed by realizing that we have different *types* of entanglement. Then a given state, even though it is entangled, may not contain some particular type of entanglement, and the measure that quantifies that type will attribute no entanglement to the state.

5.4 Concluding Remarks

In contrast to the case of pure states, the problem of mixed-state entanglement is "nondegenerate" in the sense that the various scalar and structural separability criteria are not equivalent. There is a fundamental connection between entanglement and positive maps, represented by Theorem 5.1. However, there is still a problem of turning it into an operational criterion for higher-dimensional systems. Recently [358,359] the question was reduced to a problem of investigation of the so-called "edge" PPT entangled states, as well as of the positive maps and entanglement witnesses detecting their entanglement. Some operational criteria for low-rank density matrices (and also for the multiparticle case) have been worked out in [360].

It is remarkable that the structure of entanglement reveals a discontinuity. There are two qualitatively different types of entanglement: distillable, "free" entanglement, and "bound" entanglement, which cannot be distilled. All the two-qubit entangled states are free entangled. Moreover, a free entangled state in any dimension must have some features of two-qubit entanglement. Bound entanglement is practically useless for quantum communication. However, it is not a marginal phenomenon, as the volume of the set of BE states in the set of all states for finite dimension is nonzero.

The discovery of activation of bipartite bound entanglement suggested [340] the nonadditivity of the corresponding quantum communication channels,[29] in the sense that the distillable entanglement $D(\varrho_{\mathrm{BE}} \otimes \varrho_{\mathrm{EF}})$ could exceed $D(\varrho_{\mathrm{FE}})$ for some free entangled state ϱ_{FE} and bound entangled state ϱ_{BE}. Quite recently it has been shown [339] that, in the multipartite case, two different bound entangled states, if tensored together, can make a distillable state: $D(\varrho_{\mathrm{BE}}^1 \otimes \varrho_{\mathrm{BE}}^2) > D(\varrho_{\mathrm{BE}}^1) + D(\varrho_{\mathrm{BE}}^2) = 0$. This new nonclassical effect was called *superactivation*. On the other hand, in [352] it was shown that the four-party "unlockable" bound entangled states [361] can be used for remote concentration of quantum information. It is intriguing that for bipartite systems, with the exception of the activation effect, bound entanglement is

---

[29] This could be then reformulated in terms of the so-called *binding entanglement channels* [305,355].

permanently passive. In general, there may be a qualitative difference between bipartite bound entanglement and the multipartite form. Still, in the light of recent results [362], it is quite possible that bipartite bound entanglement is also nonadditive. The very recent investigations of bound entanglement for continuous variables [363, 364] also raise analogous questions in this latter domain.

As we have seen, there is a basic connection between bound entanglement and irreversibility. As a consequence, it would be interesting to investigate some dynamical features of BE. It cannot be excluded that some systems involving BE states may reveal a nonstandard (nonexponential) decay of entanglement. In general, it seems that the role of bound entanglement in quantum communication will be negative: in fact, the existence of BE constitutes a fundamental *restriction* on entanglement processing. One can speculate that it is the ultimate restriction in the context of distillation, i.e. that it may allow one to determine the value of the distillable entanglement. Hence it seems important to develop an approach combining BE and the entanglement measures involving relative entropy. It also seems reasonable to conjecture that, in the case of general distillation processes involving the conversion of mixed states $\varrho \to \varrho'$ [66], the bound entanglement E_B never decreases[30] (i.e. $\Delta E_B \geq 0$) in optimal processes.

The irreversibility inherently connected with distillation encourages us to develop some natural *formal* analogies between mixed-state entanglement processing and phenomenological thermodynamics. The construction of a "thermodynamics of entanglement" (cf. [325, 335, 347, 365]) would be essential for a synthetic understanding of entanglement processing. Of course, progress in the above directions would require the development of various techniques of searching for bound entangled states.

One of the challenges of mixed-state entanglement theory is to determine which states are useful for quantum communication with given additional resources. In particular, we still do not know (i) which states are distillable under LOCC (i.e. which states are free entangled), and (ii) which states are distillable under *one-way* classical communication and local operations.

A promising direction for mixed-state entanglement theory is its application to the theory of quantum channel capacity, pioneered in [66]. In particular, the methods leading to upper bounds for distillable entanglement described in Sect. 5.3.7 allow one to obtain upper bounds for quantum channel capacities [101] (one of them was obtained earlier [97]). It has been shown [101] that the following hypothetical inequality

$$D_1(\varrho) \geq S(\varrho_B) - S(\varrho) , \tag{5.76}$$

---

[30] The bound entanglement can be quantified [71] as the difference between the entanglement of formation and the entanglement of distillation (defined within the original distillation scheme): $E_B = E_F - D$.

where $D_1(\varrho)$ is the *one-way* distillable entanglement,[31] would imply *equality* between the capacity of a quantum channel and the maximal rate of coherent information [366]. The latter equality would be nothing but a quantum Shannon theorem, with coherent information being the counterpart of mutual information. All the results obtained so far in the domain of quantifying entanglement indicate that the inequality is true. However, a proof of the inequality has not been found so far.

Finally, one would like to have a clear connection between entanglement and its basic manifestation – nonlocality. One can assume that free entangled states exhibit nonlocality via a distillation process [265,266]. However, the question concerning the possible nonlocality of BE states remains open (see [367–369]).

To answer the above and many other questions, one must develop the mathematical description of the structure of mixed-state entanglement. In this context, it would be especially important to push forward the mathematics of positive maps. One hopes that the exciting physics connected with mixed-state entanglement that we have presented in this contribution will stimulate progress in this domain.

---

[31] Classical messages can be sent only from Alice to Bob during distillation.

References

1. W. Tittel, G. Ribordy, N. Gisin: Phys. World, March 1998, p. 41
2. D. Deutsch, A. Ekert: Phys. World, March 1998, p. 47
3. A. Zeilinger: Phys. World, March 1998, p. 35
4. W.K. Wootters, W.H. Zurek: Nature **299**, 802 (1982)
5. E. Schrödinger: Naturwiss. **48**, 807 (1935)
6. D. Welsh: *Codes and Cryptography* (Oxford University Press, Oxford 1988)
7. D. Deutsch: Proc. R. Soc. London A **400**, 97 (1985)
8. S. Lloyd: Phys. Rev. A **61**, 010301 (2000)
9. P. Shor: In *Proceedings of the 35th Annual Symposium on the Foundations of Computer Science, 1994, Los Alamitos, California* (IEEE Computer Society Press, New York 1994) p. 124
10. D. Bouwmeester, J.-W. Pan, K. Mattle, M. Eibl, H. Weinfurter, A. Zeilinger: Nature **390**, 575 (1997); D. Boschi, S. Branca, F. De Martini, L. Hardy, S. Popescu: Phys. Rev. Lett. **80**, 1121 (1998)
11. A. Furusawa, J.L. Sørensen, S.L. Braunstein, C.A. Fuchs, H.J. Kimble, E.S. Polzik: Science **282**, 706 (1998)
12. D. Bouwmeester, A. Ekert, A. Zeilinger (eds.): *The Physics of Quantum Information*, (Springer, Berlin, Heidelberg, 2000)
13. C.A. Sackett, D. Kielpinski, B.E. King, C. Langer, V. Meyer, C.J. Myatt, M. Rowe, Q.A. Turchette, W.M. Itano, D.J. Wineland, C. Monroe: Nature **404**, 256 (2000)
14. R. Blatt: Nature **404**, 231 (2000)
15. D.G. Cory, R. Laflamme, E. Knill, L. Viola, T.F. Havel, N. Boulant, G. Boutis, E. Fortunato, S. Lloyd, R. Martinez, C. Negrevergne, M. Pravia, Y. Sharf, G. Teklemariam, Y.S. Weinstein, W.H. Zurek: LANL preprint quant-ph/0004104
16. X. Maitre, E. Hagley, G. Nogues, C. Wunderlich, P. Goy, M. Brune, J.M. Raimond, S. Haroche: Phys. Rev. Lett. **79**, 769 (1997)
17. R. Folman, P. Krüger, D. Cassettari, B. Hessmo, T. Maier, J. Schmiedmayer: Phys. Rev. Lett. **84**, 4749 (2000)
18. G.K. Brennen, C.M. Caves, P.S. Jessen, I.H. Deutsch: Phys. Rev. Lett. **82**, 1060 (1999)
19. D. Jaksch, H.J. Briegel, J.I. Cirac, C.W. Gardiner, P. Zoller: Phys. Rev. Lett. **82**, 1975 (1999)
20. J. I. Cirac and P. Zoller: Nature **404**, 579 (2000)
21. B. E. Kane: Nature **393**, 133 (1998)
22. P. M. Platzman and M. I. Dykman: Science **284**, 1967 (1999)
23. Yu. Makhlin, G. Schön, A. Shnirman: Rev. Mod. Phys. (in press)
24. A. Einstein, B. Podolsky, N. Rosen: Phys. Rev. **47**, 777 (1935)

25. J.S. Bell: Physics **1**, 195 (1964)
26. A. Peres: *Quantum Theory: Concepts and Methods* (Kluwer, Dordrecht 1995)
27. J.F. Clauser, M.A. Horne, A. Shimony, R.A. Holt: Phys. Rev. Lett. **23**, 880 (1969)
28. E.P. Wigner: Am. J. Phys. **38**, 1005 (1970)
29. F. Selleri: *Quantum Paradoxes and Physical Reality* (Kluwer, Dordrecht 1989)
30. W. Tittel, J. Brendel, H. Zbinden, N. Gisin: Phys. Rev. Lett. **81**, 3563 (1998)
31. G. Weihs, T. Jennewein, C. Simon, H. Weinfurter, A. Zeilinger: Phys. Rev. Lett. **81**, 5039 (1998)
32. J.F. Clauser, M.A. Horne: Phys. Rev. D **10**, 526 (1974); A. Garg, N.D. Mermin: Phys. Rev. D **35**, 3831 (1987)
33. E. Santos: Phys. Rev. A **46**, 3646 (1992)
34. S.F. Huelga, M. Ferrero, E. Santos: Europhys. Lett. **27**, 181 (1994); M. Freyberger, P.K. Aravind, M.A. Horne, A. Shimony: Phys. Rev. A **53**, 1232 (1996); I.C. Percival: LANL preprint quant-ph/0008097
35. E.S. Fry, T. Walther: Adv. Atom. Mol. Opt. Phys. **42**, 1 (2000)
36. E. Hagley, X. Maitre, G. Nogues, C. Wunderlich, M. Brune, J.M. Raimond, S. Haroche: Phys. Rev. Lett. **79**, 1 (1997)
37. D.M. Greenberger, M.A. Horne, A. Zeilinger: In *Bell's Theorem, Quantum Theory, and Conceptions of the Universe*, ed. by M. Kafatos (Kluwer, Dordrecht 1989) p. 73; D.M. Greenberger, M.A. Horne, A. Shimony, A. Zeilinger: Am. J. Phys. **58**, 1131 (1990); N.D. Mermin: Am. J. Phys. **58**, 731 (1990); N.D. Mermin: Phys. Today **9**, June 1990
38. J.-W. Pan, D. Bouwmeester, M. Daniell, H. Weinfurter, A. Zeilinger: Nature **403**, 515 (2000)
39. M. Hillery, V. Bužek, A. Berthiaume: Phys. Rev. A **59**, 1829 (1999)
40. R.P. Feynman: Int. J. Theor. Phys. **21**, 467 (1982)
41. S. Wiesner: Sigact News **15**, 78 (1983)
42. C.H. Bennett, G. Brassard: In *Proceedings of the IEEE International Conference on Computers, Systems and Signal Processing* (IEEE, New York 1984) p. 175
43. G.S. Vernam: J. Am. Inst. Electr. Eng. **45**, 109 (1926)
44. C.E. Shannon: Bell Syst. Tech. J. **28**, 656 (1949)
45. A. Ekert, N. Gisin, B. Huttner, H. Inamori, H. Weinfurter: In *The Physics of Quantum Information*, ed. by D. Bouwmeester, A. Ekert, A. Zeilinger (Springer, Berlin, Heidelberg, 2000) p. 15
46. A.K. Ekert: Phys. Rev. Lett. **67**, 661 (1991)
47. C.H. Bennett, F. Bessette, G. Brassard, L. Salvail, J. Smolin: J. Cryptol. **5**, 3 (1992)
48. A.K. Ekert, J.G. Rarity, P.R. Tapster, G.M. Palma: Phys. Rev. Lett. **69**, 1293 (1992)
49. A. Muller, J. Breguet, N. Gisin: Europhys. Lett. **23**, 383 (1993)
50. G. Ribordy, J.D. Gautier, N. Gisin, O. Guinnard, H. Zbinden: Electron. Lett. **34**, 2116 (1998)
51. C.H. Bennett, S.J. Wiesner: Phys. Rev. Lett. **69**, 2881 (1992)
52. C.H. Bennett, G. Brassard, C. Crepeau, R. Josza, A. Peres, W.K. Wootters: Phys. Rev. Lett. **70**, 1895 (1993)
53. M. Zukowski, A. Zeilinger, M.A. Horne, A. Ekert: Phys. Rev. Lett. **71**, 4287 (1993)

54. P. Benioff: Phys. Rev. Lett. **48**, 1581 (1982)
55. C.H. Bennett: IBM J. Res. Dev. **17**, 525 (1973)
56. D. Deutsch, R. Josza: Proc. R. Soc. London A **439**, 553 (1992)
57. D. Simon: In *Proceedings of the 35th Annual Symposium on the Foundations of Computer Science, 1994, Los Alamitos, California* (IEEE Computer Society Press, New York 1994) p. 116
58. L. Grover: In *Proceedings of the 28th Annual Symposium on the Theory of Computing, 1996, Philadelphia, Pennsylvania* (ACM Press, New York 1996) p. 212; Phys. Rev. Lett. **79**, 325 (1997)
59. J.I. Cirac, P. Zoller: Phys. Rev. Lett. **74**, 4091 (1995)
60. T. Pellizzari, S.A. Gardiner, J.I. Cirac, P. Zoller: Phys. Rev. Lett. **75**, 3788 (1995)
61. H. Mabuchi, P. Zoller: Phys. Rev. Lett. **76**, 3108 (1996)
62. A. Steane: Phys. Rev. Lett. **77**, 793 (1996)
63. R. Laflamme, C. Miquel, J.P. Paz, W.H. Zurek: Phys. Rev. Lett. **77**, 198 (1996)
64. A. Ekert, C. Macchiavello: Phys. Rev. Lett. **77**, 2585 (1996)
65. D. Gottesman: Phys. Rev. A **54**, 1862 (1996)
66. C.H. Bennett, D.P. DiVincenzo, J.A. Smolin, W.K. Wootters: Phys. Rev. A **54**, 3824 (1996)
67. L.B. Levitin: In *Proceedings of the 4th All-Union Conference on Information Transmission and Coding Theory* (Tashkent 1969) p. 111
68. A.S. Holevo: Probl. Peredachi Inform. **9**, 3 (1973) [Probl. Inf. Trans. (USSR) **9**, 177 (1973)]
69. B. Schumacher: Phys. Rev. A **51**, 2738 (1995)
70. S. Braunstein, H.J. Kimble: Phys. Rev. Lett. **80**, 869 (1998)
71. M. Horodecki, P. Horodecki, R. Horodecki: Phys. Rev. Lett. **80**, 5239 (1998)
72. J.S. Clauser, A. Shimony: Rep. Prog. Phys. **41**, 1881 (1978)
73. A. Aspect, P. Grangier, G. Roger: Phys. Rev. Lett **47**, 460 (1981)
74. R. Derka, V. Bužek, A.K. Ekert: Phys. Rev. Lett. **80**, 1571 (1998)
75. N. Gisin, S. Massar: Phys. Rev. Lett. **79**, 2153 (1997)
76. M. Keyl, R.F. Werner: J. Math. Phys. **40**, 3283 (1999)
77. V. Bužek, M. Hillery, R.F. Werner: Phys. Rev. A **60**, R2626 (1999)
78. J.I. Cirac, A.K. Ekert, C. Macchiavello: Phys. Rev. Lett. **82**, 4344 (1999)
79. M. Keyl, R.F. Werner: LANL preprint quant-ph/9910124 (to appear in Ann. Henri Poincaré)
80. P. Shor: SIAM J. Comput. **26**, 1484 (1997) (LANL preprint quant-ph/9508027)
81. A.S. Holevo, R.F. Werner: Phys. Rev. A **63**, 032312 (2001) (LANL preprint quant-ph/9912067)
82. A. Calderbank, P. Shor: Phys. Rev. A **54**, 1098 (1996)
83. S.J. Summers, R.F. Werner: Commun. Math. Phys. **110**, 247 (1987)
84. H. Halvorson, R. Clifton: J. Math. Phys. **41**, 1711 (2000)
85. O. Bratteli, D.W. Robinson: *Operator Algebras and Quantum Statistical Mechanics*, vol. 1 (Springer, Berlin 1979)
86. E.B. Davies: *Quantum Theory of Open Systems* (Academic Press, New York 1976)
87. J. von Neumann: *Mathematical Foundations of Quantum Mechanics* (Princeton University Press, Princeton 1955)

88. K. Kraus: *States, Effects, and Operations*, Lecture Notes in Physics, vol. 190 (Springer, Berlin, Heidelberg 1983)
89. W.F. Stinespring: Proc. Am. Math. Soc. **6**, 211 (1955)
90. W. Arveson:Acta Math. **123**, 141 (1969)
91. V.I. Paulsen: *Completely Bounded Maps and Dilations* (Longman, Harlow 1986)
92. C.E. Shannon: Bell Syst. Tech. J. **27**, 379 (1948); Bell Syst. Tech. J. **27**, 623 (1948)
93. B.W. Schumacher: Phys. Rev. A **54**, 2614 (1996)
94. H. Barnum, E. Knill, M.A. Nielsen: IEEE Trans. Informat. Theory **46**, 1317 (2000) (LANL preprint quant-ph/9809010)
95. M. Ohya, D. Petz: *Quantum Entropy and its Use* (Springer, Berlin, Heidelberg 1993)
96. A.S. Holevo: Tamagawa Univ. Res. Rev., No. 4 (1998) (LANL preprint quant-ph/9809023)
97. H. Barnum, M. Nielsen, B. Schumacher: Phys. Rev. A **57**, 4153 (1998)
98. N.J. Cerf, C. Adami: Phys. Rev. Lett. **79**, 5194 (1997)
99. D.P. DiVincenzo, P.W. Shor, J.A. Smolin: Phys. Rev. A **57**, 830 (1998)
100. S. Lloyd: Phys. Rev. A **56**, 1613 (1997)
101. M. Horodecki, P. Horodecki, R. Horodecki: Phys. Rev. Lett. **85**, 433 (2000)
102. R.F. Werner: LANL preprint quant-ph/0003070 (to appear in J. Phys. A)
103. T. Beth, D. Jungnickel, H. Lenz: *Design Theory*, vol. 1, 2nd edn. (Cambridge University Press, Cambridge 1999)
104. T. Beth, D. Jungnickel, H. Lenz: *Design Theory*, vol. 2, 2nd edn. (Cambridge University Press, Cambridge 1999)
105. C.H. Bennett: Phys. Today **48**, No. 10, 24 (1995)
106. *Introduction to Quantum Computation and Information*, ed. by H.-K. Lo, S. Popescu, T.P. Spiller (World Scientific, Singapore 1999)
107. N. Bohr: Phys. Rev. **48**, 696 (1935)
108. J. von Neumann: *Mathematische Grundlagen der Quantenmechanik* (Springer, Berlin 1932)
109. J.S. Bell: Physics **1**, 195 (1964); Rev. Mod. Phys. **38**, 447 (1966); In *Foundations of Quantum Mechanics*, ed. B. D'Espagnat (Academic Press, New York 1971) p. 171; *Speakable and Unspeakable in Quantum Mechanics* (Cambridge University Press, Cambridge 1987)
110. P.M. Pearle: Phys. Rev. D **2**, 1418 (1970)
111. P.G. Kwiat, K. Mattle, H. Weinfurter, A. Zeilinger, A.V. Sergienko, Y.H. Shih: Phys. Rev. Lett **75**, 4337 (1995)
112. H. Weinfurter: Europhys. Lett. **25**, 559 (1994); A. Zeilinger, H.J. Bernstein, M.A. Horne: J. Mod. Opt., **41**, 2375 (1994)
113. T. Jennewein, G. Weihs, C. Simon, H. Weinfurter, A. Zeilinger: Phys. Rev. Lett. **54**, 4729 (2000)
114. K. Mattle, H. Weinfurter, P.G. Kwiat, A. Zeilinger: Phys. Rev. Lett. **76**, 4656 (1996)
115. J.-W. Pan, D. Bouwmeester, H. Weinfurter, A. Zeilinger: Phys. Rev. Lett. **80**, 3891 (1998)
116. A. Peres: *Quantum Theory: Concepts and Methods* (Kluwer, Dordrecht 1993); A. Peres: In *Proceedings of the Symposium on Fundamental Problems in Quantum Physics*, Oviedo 1996 (Kluwer Academic, Dordrecht 1997);

P. Horodecki: Phys. Lett. A **232**, 333 (1997); M. Horodecki, P. Horodecki, R. Horodecki: Phys. Rev. Lett. **80**, 5239 (1998); N. Linden, S. Popescu: Fortsch. Phys. **46**, 567 (1998)

117. D.M. Greenberger, M.A. Horne, A. Zeilinger: Phys. Today, August 1993, p. 22; G. Krenn, A. Zeilinger: Phys. Rev. A **54**, 1793 (1996)
118. M. Zukowski, A. Zeilinger, M.A. Horne: Phys. Rev. A **55**, 2564 (1997); A. Zeilinger, M. Zukowski, M.A. Horne, H.J. Bernstein, D.M. Greenberger: In *Quantum Interferometry*, ed. by F. DeMartini, A. Zeilinger (World Scientific, Singapore 1994) p. 134
119. C.H. Bennett, G. Brassard, S. Popescu, B. Schumacher, J.A. Smolin, W.K. Wootters: Phys. Rev. Lett. **76**, 722 (1996); C.H. Bennett, D. DiVincenzo, J.A. Smolin, W.K. Wootters: Phys. Rev. A **54**, 3824 (1996); D. Deutsch, A. Ekert, R. Josza, C. Macchiavello, S. Popescu, A. Sanpera: Phys. Rev. Lett. **77**, 2818 (1996)
120. C.H. Bennett, D.P. DiVincenzo, C.A. Fuchs, T. Mor, E. Rains, P.W. Shor, J.A. Smolin, W.K. Wootters: Phys. Rev. A **59**, 1070 (1999)
121. S.L. Braunstein, A. Mann, M. Revzen: Phys. Rev. Lett. **68**, 3259 (1992)
122. C.H. Bennett, G. Brassard, A. Ekert: Sci. Am. October 1992, p. 26
123. C.H. Bennett, G. Brassard, N.D. Mermin: Phys. Rev. Lett. **68**, 557 (1992)
124. L.M. Krauss: *Physics of Star Trek* (Basic Books, New York 1995)
125. M. Zukowski, A. Zeilinger, M.A. Horne, A.K. Ekert: Phys. Rev. Lett. **71**, 4287 (1993); S. Bose, V. Vedral, P.L. Knight: Phys. Rev. A **57**, 822 (1998)
126. L. Vaidman: Phys. Rev. A **49**, 1473 (1994); S.L. Braunstein, H.J. Kimble: Phys. Rev. Lett. **80**, 869 (1998); T.C. Ralph, P.K. Lam: Phys. Rev. Lett. **81**, 5668 (1998)
127. S. Popescu: LANL preprint quant-ph/9501020
128. Q.A. Turchette, C.S. Wood, B.E. King, C.J. Myatt, D. Leibfried, W.M. Itano, C. Monroe, D.J. Wineland: Phys. Rev. Lett. **81**, 3631 (1998)
129. C.J. Hood, M.S. Chapman, T.W. Lynn, H.J. Kimble: Phys. Rev. Lett. **80**, 4157 (1998)
130. A. Imamoglu, H. Schmidt, G. Woods, M. Deutsch: Phys. Rev. Lett. **79**, 1467 (1997); M.J. Werner, A. Imamoglu: Phys. Rev. A **61**, 011801 (2000) (LANL preprint quant-ph/9902005)
131. J.D. Franson: Phys. Rev. Lett. **78**, 3852 (1998)
132. C.S. Wu, I. Shaknov: Phys. Rev. **77**, 136 (1950)
133. S.J. Freedman, J.S. Clauser: Phys. Rev. Lett. **28**, 938 (1972); A. Aspect, J. Dalibard, G. Roger: Phys. Rev. Lett. **47**, 1804 (1982)
134. R.W. Boyd: *Nonlinear Optics* (Academic Press, San Diego 1992)
135. R. Gosh, C.K. Hong, Z.Y. Ou, L. Mandel: Phys. Rev. A **34**, 3962 (1986); A.E. Steinberg, P.G. Kwiat, R.Y. Chiao: In *Atomic, Molecular and Optical Physics Handbook*, ed. by G. Drake (AIP Press, New York 1996), Chap. 77, p. 901
136. Y.H. Shih, C.O. Alley: Phys. Rev. Lett. **61**, 2921 (1988)
137. M.D. Reid, D.F. Walls: Phys. Rev. A **34**, 1260 (1986)
138. M. Michler, K. Mattle, H. Weinfurter, A. Zeilinger: Phys. Rev. A **53**, R1209 (1996)
139. J.D. Franson: Phys. Rev. Lett. **62**, 2205 (1989)
140. P.G. Kwiat, W.A. Vareka, C.K. Hong, C.K. Nathel, R.Y. Chiao: Phys. Rev. A **41**, 2910 (1990); Z.Y. Ou, X.Y. Zou, L.J. Wang, L. Mandel: Phys. Rev. Lett.

      65, 321 (1990); J. Brendel, E. Mohler, W. Martienssen: Phys. Rev. Lett. **66**, 1142 (1991); W. Tittel, J. Brendel, H. Zbinden, N. Gisin: Phys. Rev. Lett. **81**, 3563 (1998)
141. S. Popescu, L. Hardy, M. Zukowski: Phys. Rev. A **56**, R4353 (1997); S. Aerts, P.G. Kwiat, J.-Å. Larsson, M. Zukowski: Phys. Rev. Lett. **83**, 2872 (1999)
142. A.K. Ekert, J.G. Rarity, P.R. Tapster, G.M. Palma: Phys. Rev. Lett. **69**, 1293 (1992); P.D. Townsend, J.G. Rarity, P.R. Tapster: Electron. Lett. **29**, 634 (1993)
143. M.A. Horne, A. Shimony, A. Zeilinger: Phys. Rev. Lett. **62**, 2209 (1989)
144. J.G. Rarity, P.R. Tapster: Phys. Rev. Lett. **64**, 2495 (1990); Z.Y. Ou, X.Y. Zou, L.J. Wang, L. Mandel: Phys. Rev. Lett. **65**, 321 (1990)
145. J. Brendel, N. Gisin, W. Tittel, H. Zbinden: Phys. Rev. Lett. **82**, 2594 (1999)
146. P.G. Kwiat: Ph.D. Thesis, University of California at Berkeley, 1993; A. Garuccio: Ann. N. Y. Acad. Sci. **755**, 632 (1995)
147. M.H. Rubin, D.N. Klyshko, Y.H. Shih, A.V. Sergienko: Phys. Rev. A **50**, 5122 (1994)
148. P.G. Kwiat, E. Waks, A.G. White, I. Appelbaum, P.H. Eberhard: Phys. Rev. A **60**, R773 (1999)
149. M. Oberparleiter, H. Weinfurter: Opt. Commun. **183**, 133 (2000)
150. A. Zeilinger, H.J. Bernstein, M.A. Horne: J. Mod. Opt. **41**, 2375 (1994)
151. S.L. Braunstein, A. Mann: Phys. Rev. A **51**, R1727 (1995); M. Michler, K. Mattle, H. Weinfurter, A. Zeilinger: Phys. Rev. A **53**, R1209 (1996)
152. R. Loudon: In *Coherence and Quantum Optics VI*, ed. by J.K. Eberly (Plenum, New York 1990) p. 703
153. C.K. Hong, Z.Y. Ou, L. Mandel: Phys. Rev. Lett. **59**, 2044 (1987)
154. H. Fearn, R. Loudon: J. Opt. Soc. Am. **6**, 917 (1989); J.G. Rarity, P.R. Tapster, E. Jakeman, T. Larchuk, R.A. Campos, M.C. Teich, B.E.A. Saleh: Phys. Rev. Lett. **65**, 1348 (1990); R.A. Campos, M.C. Teich, B.E.A. Saleh: Phys. Rev. A **42**, 4127 (1990); K. Mattle, M. Michler, H. Weinfurter, A. Zeilinger, M. Zukowski: Appl. Phys. B **60**, S111 (1995)
155. L. Vaidman, N. Yoran: Phys. Rev. A **59**, 116 (1999); N. Lütkenhaus, J. Calsamiglia, K.-A. Suominen: Phys. Rev. A **59**, 3295 (1999)
156. P.G. Kwiat, H. Weinfurter: Phys. Rev. A **58**, R2623 (1998)
157. M. Zukowski, A. Zeilinger, H. Weinfurter: Ann. N. Y. Acad. Sci. **755**, 91 (1995); J.G. Rarity: Ann. N. Y. Acad. Science **755**, 624 (1995)
158. Z.Y. Ou, Y.L. Lu: Phys. Rev. Lett. **83**, 2556 (1999)
159. P.D. Townsend, J.G. Rarity, P.R. Tapster: Electron. Lett. **29**, 634 (1993); P.R. Tapster, J.G. Rarity, P.C.M. Owens: Phys. Rev. Lett. **73**, 1823 (1994)
160. D.S. Naik, C.G. Peterson, A.G. White, A.J. Berglund, P.G. Kwiat: Phys. Rev. Lett. **84**, 4733 (2000); W. Tittel, J. Brendel, H. Zbinden, N. Gisin: Phys. Rev. Lett. **84**, 4737 (2000); G. Ribordy, J. Brendel, J.-D. Gautier, N. Gisin, H. Zbinden: LANL preprint quant-ph/0008039
161. T. Jennewein, U. Achleitner, G. Weihs, H. Weinfurter, A. Zeilinger: Rev. Sci. Instrum. **41**, 1675 (2000)
162. H.-J. Briegel, W. Dür, J.I. Cirac, P. Zoller: Phys. Rev. Lett. **81**, 5932 (1998); W. Dür, H.-J. Briegel, J.I. Cirac, P. Zoller: Phys.Rev. A **59**, 169 (1999)
163. L. Grover: LANL preprint quant-ph/9704012; J.I. Cirac, A. Ekert, S. Huelga, C. Macchiavello: Phys. Rev. A **59**, 4249 (1999); J.I. Cirac, P. Zoller, H.J. Kimble, H. Mabuchi: Phys. Rev. Lett. **78**, 3221 (1997)

164. R. Landauer: Phys. Lett. A **217**, 188 (1996); R. Landauer: Proc. R. Soc. London A **454**, 305 (1998); S. Haroche, J.M. Raimond: Phys. Today, August 1996, p. 51
165. L. Goldenberg, L. Vaidman, S. Wiesner: Phys. Rev. Lett. **82**, 3356 (1999)
166. J. Eisert, M. Wilkens, M. Lewenstein: Phys. Rev. Lett. **83**, 3077 (1999)
167. L. Hardy, W. van Dam: Phys. Rev. A **59**, 2635 (1999)
168. R. Cleve, H. Buhrman: Phys. Rev. A **56**, 1201 (1997); R. Cleve, W. van Dam, M. Nielsen, A. Tapp: In *Proceedings of the 1st NASA International Conference on Quantum Computing and Quantum Communication* (Springer, Berlin, Heidelberg 1998)
169. V. Bužek, M. Hillery: Phys. Rev. A **54**, 1844 (1996); N. Gisin, S. Massar: Phys. Rev. Lett. **79**, 2153 (1997)
170. D. Bouwmeester, J.-W. Pan, M. Daniell, H. Weinfurter, A. Zeilinger: Phys. Rev. Lett. **82**, 1345 (1999)
171. T. Beth: *Verfahren der Schnellen Fouriertransformation* (Teubner, Stuttgart 1984)
172. A. Barenco, C.H. Bennett, R. Cleve, D.P. DiVincenzo, N. Margolus, P. Shor, T. Sleator, J.A. Smolin, H. Weinfurter: Phys. Rev. A **52**, 3457 (1995)
173. P. Bürgisser, M. Clausen, A. Shokrollahi: *Algebraic Complexity Theory*, Grundlehren der mathematischen Wissenschaften, vol. 315 (Springer, Berlin, Heidelberg 1997)
174. M. Clausen, U. Baum: *Fast Fourier Transforms* (BI-Wissenschaftsverlag, Mannheim 1993)
175. D.P. DiVincenzo: Phys. Rev. A **51**, 1015 (1995)
176. A. Barenco: Proc. R. Soc. London A **449**, 679 (1995)
177. D. Deutsch, A. Barenco, A. Ekert: Proc. R. Soc. London A **449**, 669 (1995)
178. E. Knill: Technical report, Los Alamos National Laboratory, 1995; LANL preprint quant-ph/9508006
179. E. Knill: Technical report, Los Alamos National Laboratory, 1995; LANL preprint quant-ph/9508007
180. A.Yu. Kitaev: Russ. Math. Surveys **52**, 1191 (1997)
181. C. Moore, M. Nilsson: Technical report, Los Alamos National Laboratory, 1997; LANL preprint quant-ph/9804034
182. D. Jungnickel: *Finite Fields: Structure and Arithmetics* (BI-Wissenschaftsverlag, Mannheim 1993)
183. I. Wegener: *The Complexity of Boolean Functions* (Teubner, Stuttgart 1987)
184. F. MacWilliams, N. Sloane: *The Theory of Error-Correcting Codes* (North-Holland, Amsterdam 1977)
185. T. Toffoli: In *Proceedings of ICALP'80* (Springer, Berlin, Heidelberg 1980) p. 632
186. P. Fuhrmann: *A Polynomial Approach to Linear Algebra* (Springer, Berlin, Heidelberg 1996)
187. A.Yu. Kitaev: Technical report, Los Alamos National Laboratory, 1995; LANL preprint quant-ph/9511026
188. M. Grassl, W. Geiselmann, T. Beth: In *Proceedings of Applied Algebra, Algebraic Algorithms and Error-Correcting Codes (AAECC-13)*, Lecture Notes in Computer Science, vol. 1719 (Springer, Berlin, Heidelberg 1999) p. 231
189. G.H. Golub, C.F. van Loan: *Matrix Computations*, 3rd edn. (John Hopkins University Press, Baltimore 1996)

190. V. Vedral, A. Barenco, A. Ekert: Phys. Rev. A **54**, 147 (1996)
191. D. Beckman, A.N. Chari, S. Devabhaktuni, J. Preskill: Phys. Rev. A **54**, 1034 (1996)
192. A. Barenco, A. Berthiaume, D. Deutsch, A. Ekert, R. Jozsa, C. Macchiavello: SIAM J. Comput. **26**, 1541 (1997)
193. D. Knuth: *The Art of Computer Programming*, vol. 3 (Addison-Wesley, Reading, MA 1973)
194. A. Turing: Proc. London Math. Soc. **42**, 230 (1936); correction, Proc. London Math. Soc. **43**, 554 (1937)
195. J. van Leeuwen (ed.): *Handbook of Theoretical Computer Science*, vol. A: *Algorithms and Complexity* (Elsevier, Amsterdam 1990)
196. A. Church: Am. J. Math. **58**, 345 (1936)
197. S. Kleene: *Introduction to Metamathematics*, Bibliotheca Mathematica (North-Holland, Amsterdam 1952)
198. C.H. Bennett: IBM J. Res. Dev. **32**, 16 (1988)
199. C.H. Bennett: SIAM J. Comput. **18**, 766 (1989)
200. E. Bernstein, U. Vazirani: SIAM J. Comput. **26**, 1411 (1997)
201. L. Blum, M. Shub, S. Smale: Bull. Am. Math. Soc. **21**, 1 (1989)
202. L. Adleman, J. Demarrais, M. Huang: SIAM J. Comput. **26**, 1524 (1997)
203. M. Rötteler, M. Püschel, T. Beth: In *Proceedings of the Workshop on Physics and Computer Science*, ed. by W. Kluge (DPG-Frühjahrstagung, Heidelberg 1999) p. 31
204. A. Yao: In *Proceedings of the 34th Annual Symposium on the Foundations of Computer Science* (IEEE Computer Society Press, Los Alamitos 1993) p. 352
205. A. Berthiaume: In *Complexity Theory Retrospective II*, chapter on "Quantum computation" (Springer, Berlin, Heidelberg 1996) p. 23
206. G. Brassard, P. Høyer: In *Proceedings of the Fifth Israeli Symposium on Theory of Computing and Systems* (ISTCS, IEEE Computer Society Press, New York 1997) p. 12
207. H. Cohen: *A Course in Computational Algebraic Number Theory*, 3rd edn. (Springer, Berlin, Heidelberg 1996)
208. D.F. Elliott, K.R. Rao: *Fast Transforms — Algorithms, Analyses, Applications* (Academic Press, New York 1982)
209. J.W. Cooley, J.W. Tukey: Math. Comput. **19**, 297 (1965)
210. R. Griffiths, C. Niu: Phys. Rev. Lett. **76**, 3228 (1996)
211. S. Lang: *Algebra* (Addison-Wesley, Reading, MA 1993)
212. N. Jacobson: *Basic Algebra II* (Freeman, New York 1989)
213. J.P. Serre: *Linear Representations of Finite Groups* (Springer, New York 1977)
214. P. Davis: *Circulant Matrices* (Wiley Interscience, New York 1979)
215. L. Grover: Phys. Rev. Lett. **80**, 4329 (1998)
216. A. VanderLugt: *Optical Signal Processing* (Wiley, New York 1992)
217. M. Schmid, R. Steinwandt, J. Müller-Quade, M. Rötteler, T. Beth: Lin. Algebra Appl. **306**, 131 (2000)
218. G. Brassard, P. Høyer, A. Tapp: In *Automata, Languages and Programming*, Lecture Notes in Computer Science, vol. 1443, ed. by K. Larsen (Springer, Berlin. Heidelberg 1998) p. 820
219. E. Biham, O. Biham, D. Biron, M. Grassl, D. Lidar: Phys. Rev. A **60**, 2742 (1999)

220. G. Brassard, P. Høyer, A. Tapp: ACM SIGACT News (Cryptol. Col.) **28**, 14 (1997)
221. H. Buhrman, R. Cleve, A. Widgerson: In *Proceedings of the Symposium on Theory of Computing (STOC)*, Dallas, Texas, 1998
222. G.L. Miller: J. Comp. Sys. Sci. **13**, 300 (1976)
223. T. Beth, H. Aagedal: In *7. Arbeitsgespräch der Fachgruppe Physik, Informatik, Informationstechnik*, ed. by H. Hofmann, *Optik in der Rechentechnik* (1996) p. 49
224. G. Hardy, E. Wright: *An Introduction to the Theory of Numbers*, 5th edn. (Clarendon Press, Oxford 1979)
225. D. Maslen, D. Rockmore: In *Proceedings of the DIMACS Workshop in Groups and Computation*, vol. 28 (AMS Press, Providence, 1997) p. 182
226. M. Püschel: *Konstruktive Darstellungstheorie und Algorithmengenerierung*, Dissertation, University of Karlsruhe (1998)
227. W.C. Curtis, I. Reiner: *Methods of Representation Theory*, vol. 1. (Wiley Interscience, New York 1981)
228. M. Püschel, M. Rötteler, T. Beth: In *Proceedings of Applied Algebra, Algebraic Algorithms and Error-Correcting Codes (AAECC-13)*, Lecture Notes in Computer Science, vol. 1719 (Springer, Berlin, Heidelberg 1999) p. 148
229. L. Viola, E. Knill, S. Lloyd: Phys. Rev. Lett. **82**, 2417 (1999)
230. P. Zanardi: Phys. Lett. A **57**, 3276 (1998)
231. P. Zanardi: Phys. Rev. A **60**, 729 (1999)
232. P. Zanardi, M. Rasetti: Phys. Rev. Lett. **79**, 3306 (1997)
233. D. Rockmore: In *Proceedings of the DIMACS Workshop in Groups and Computation*, vol. 28 (AMS Press, Providence 1997) p. 329
234. T. Minkwitz: Applic. Algebra Eng. Commun. Comput. **7**, 391 (1996)
235. B. Huppert: *Endliche Gruppen*, vol. 1 (Springer, Berlin, Heidelberg 1983)
236. P. Høyer: Technical report, Los Alamos National Laboratory, 1997; LANL preprint quant-ph/9702028
237. R. Beals: In *Proceedings of the Symposium on Theory of Computing (STOC)*, El Paso, Texas, 1997
238. G. James, A. Kerber: *The Representation Theory of the Symmetric Group* (Addison-Wesley, Reading, MA 1981)
239. M. Ettinger, P. Høyer: In *Proceedings of the Annual Symposium on Theoretical Aspects of Computer Science (STACS)*, Lecture Notes in Computer Science, vol. 1563 (Springer, Berlin, Heidelberg 1999) p. 478
240. R. Jozsa: Proc. R. Soc. London A **454**, 323 (1998)
241. J. Köbler, U. Schöning, J. Toran: *The Graph Isomorphism Problem* (Birkhäuser, Basel 1993)
242. M. Ettinger, P. Høyer: Technical report, Los Alamos National Laboratory, 1999; LANL preprint quant-ph/9901029
243. M. Rötteler, T. Beth: Technical report, Los Alamos National Laboratory, 1998; LANL preprint quant-ph/9812070
244. A. Fijany, C.P. Williams: In *Proceedings of NASA Conference QCQC 98*, Lecture Notes in Computer Science, vol. 1509 (Springer, Berlin, Heidelberg 1998) p. 10
245. A. Klappenecker, T. Beth, M. Grassl: In *Fachtagung Informations- und Mikrosystemtechnik*, Magdeburg (1998) p. 145
246. A. Klappenecker: Technical report, Los Alamos National Laboratory, 1999; LANL preprint quant-ph/9909014

247. K.R. Rao P. Yip: *Discrete Cosine Transform: Algorithms, Advantages, and Applications* (Academic Press, Boston 1990)
248. D. Coppersmith: Technical Report RC 19642, IBM Research Division, 1994
249. T. Beth: Theor. Comput. Sci. **51**, 331 (1987)
250. M. Rötteler, T. Beth: In *Proceedings of the 10th International Symposium on Theoretical Electrical Engineering (ISTET'99)*, Magdeburg (1999) p. 85
251. R.J. McEliece: *The Theory of Information and Coding* (Addison-Wesley, Reading, MA 1977)
252. T. Beth: Designs Codes Cryptogr. **3**, 199 (1993)
253. M. Karpovsky: *Finite Orthogonal Series in the Design of Digital Devices* (Wiley, New York 1976)
254. M. Harwit, N. Sloane: *Hadamard Transform Optics* (Academic Press, New York 1979)
255. T. Beth, M. Grassl: Fortschr. Phys. **46**, 459 (1998)
256. J. Schlienz, G. Mahler: Phys. Lett. A **224**, 39 (1996)
257. B. d'Espagnat: *Conceptual Foundations of Quantum Mechanics* (Benjamin, Reading, MA 1976)
258. D. Bohm: Phys. Rev. **85**, 166 (1952)
259. W. Heitler, F. London: Z. Phys. **44**, 455 (1927)
260. J. Gruska: *Quantum Computing* (McGraw-Hill, London 1999)
261. C. Bennett, G. Brassard, C. Crepeau, R. Jozsa, A. Peres, W.K. Wootters: Phys. Rev. Lett. **70**, 1895 (1993)
262. D. Bouwmeester, J.-W. Pan, K. Mattle, M. Elbl, H. Weinfurter, A. Zeilinger: Nature **390**, 575 (1997); D. Boschi, S. Brance, F. De Martini, L. Hardy, S. Popescu, Phys. Rev. Lett. **80**, 1121 (1998); A. Furusawa, J.L. Sørensen, S.L. Braunstein, C.A. Fuchs, H.J. Kimble, E.S. Polzik: Science **282**, 706 (1998); M.A. Nielsen, E. Knill, R. Laflamme: Nature **396**, 52 (1998)
263. R.F. Werner: Phys. Rev. A **40**, 4277 (1989)
264. S. Popescu: Phys. Rev. Lett. **72**, 797 (1994)
265. S. Popescu: Phys. Rev. Lett. **74**, 2619 (1995)
266. C.H. Bennett, G. Brassard, S. Popescu, B. Schumacher, J. Smolin, W.K. Wootters: Phys. Rev. Lett. **76**, 722 (1996)
267. M. Horodecki, P. Horodecki, R. Horodecki: Phys. Lett. A **223**, 1 (1996)
268. V. Vedral, M.B. Plenio, M.A. Rippin, P.L. Knight: Phys. Rev. Lett. **78**, 2275 (1997)
269. V. Vedral, M. Plenio: Phys. Rev. A **57**, 1619 (1998)
270. G. Vidal, R. Tarrach: Phys. Rev. A, **59**, 141 (1999)
271. G. Vidal: J. Mod. Opt. **47**, 355 (2000)
272. M. Murao, M.B. Plenio, S. Popescu, V. Vedral, P.L. Knight: Phys. Rev. A **57**, R4075 (1998); W. Dür, J.I. Cirac, R. Tarrach: Phys. Rev. Lett. **83**, 3562 (1999); N. Linden, S. Popescu, A. Sudbery: Phys. Rev. Lett. **83**, 243 (1999)
273. C.H. Bennett, D. DiVincenzo, T. Mor, P. Shor, J. Smolin, B. Terhal: Phys. Rev. Lett. **82**, 5385 (1999)
274. P. Horodecki: Phys. Lett. A **232**, 233 (1997)
275. E. Schmidt: Math. Ann. **63**, 433 (1907)
276. N. Gisin: Phys. Lett. A **154**, 201 (1991)
277. R. Horodecki, P. Horodecki, M. Horodecki: Phys. Lett. A **210**, 377 (1996); R. Horodecki, M. Horodecki: Phys. Rev. A **54**, 1838 (1996)
278. R. Horodecki, P. Horodecki, M. Horodecki: Phys. Lett. A **200**, 340 (1995)

279. P. Horodecki, R. Horodecki: Phys. Rev. Lett. **76**, 2196 (1996)
280. R. Horodecki, P. Horodecki: Phys. Lett. A **194**, 147 (1994)
281. M. Horodecki, P. Horodecki: Phys. Rev. A **59**, 4206 (1999)
282. P. Horodecki, J.A. Smolin, B. Terhal, A.V. Thapliyal: LANL preprint quant-ph/99100122
283. M. Żukowski, R. Horodecki, M. Horodecki, P. Horodecki: Phys. Rev. A **58**, 1964 (1998)
284. A. Peres: Phys. Rev. Lett. **77**, 1413 (1996)
285. L.-M. Duan, G. Giedke, J.I. Cirac, P. Zoller: Phys. Rev. Lett. **84**, 2722 (2000)
286. R. Simon: Phys. Rev. Lett. **84**, 2726 (2000)
287. M.D. Choi: Lin. Algebra Appl. **12**, 95 (1975)
288. E. Størmer: Acta. Math. **110**, 233 (1963)
289. S.L. Woronowicz: Rep. Math. Phys. **10**, 165 (1976)
290. B. Terhal: *Quantum Algorithms and Quantum Entanglement*, Ph.D. Thesis, University of Amsterdam, Amsterdam (1999)
291. A. Jamiołkowski: Rep. Math. Phys. **3**, 275 (1972)
292. H. Osaka: Lin. Algebra Appl. **135**, 73 (1991)
293. V. Bužek, V. Vedral, M.B. Plenio, P.L. Knight, M. Hillery: Phys. Rev. A **55**, 3327 (1997); S. Bandyopadhyay, G. Kar: Phys. Rev. A **60**, 3296 (1999)
294. V. Bužek, S.L. Braunstein, M. Hillery, D. Bruss: Phys. Rev. A **56**, 3446 (1997); V. Bužek, M. Hillery: Phys. Rev. Lett. **81**, 5003 (1998)
295. S. Bandyopadhay, G. Kar, A. Roy: Phys. Lett. A, **258**, 205 (1999)
296. J.F. Poyatos, J.I. Cirac, P. Zoller: Phys. Rev. Lett. **78**, 390 (1997)
297. K. Życzkowski, P. Horodecki, A. Sanpera, M. Lewenstein: Phys. Rev. A **58**, 883 (1998)
298. K. Życzkowski: Phys. Rev. A **60**, 3496 (1999)
299. A. Sanpera, R. Tarrach, G. Vidal: Phys. Rev. A **58**, 826 (1998)
300. D. Bruss: Phys. Rev. A **60**, 4344 (1999)
301. S. Hill, W.K. Wootters: Phys. Rev. Lett. **78**, 5022 (1997); W.K. Wooters: Phys. Rev. Lett. **80**, 2245 (1998)
302. J. Eisert, M. Plenio: J. Mod. Opt. **46**, 1 (1999)
303. M. Horodecki, P. Horodecki, R. Horodecki: Phys. Rev. Lett. **78**, 574 (1997)
304. E. Størmer: Proc. Am. Math. Soc. **86**, 402 (1982)
305. D.P. DiVincenzo, T. Mor, P. Shor, J.A. Smolin, B. Terhal: LANL preprint quant-ph/9908070
306. D. Bruss, A. Peres: Phys. Rev. A **61**, 030301R (2000)
307. M. Lewenstein, A. Sanpera: Phys. Rev. Lett. **80**, 2261 (1998)
308. B. Terhal: Lin. Alg. Appl. **323**, 61 (2000) (LANL preprint quant-ph/9810091)
309. N. Cerf, C. Adami, R.M. Gingrich: Phys. Rev. **60**, 898 (1999)
310. D. Dür, J.I. Cirac, M. Lewenstein, D. Bruss: Phys. Rev. A **61**, 062313 (2000) (LANL preprint quant-ph/9910022)
311. E.M. Rains: Phys. Rev. A **60**, 179 (1999)
312. P. Horodecki: *Conditions for quantum separability of mixed states and distillation of quantum entanglement*, Ph.D. Thesis [in Polish], Politechnika Gdańska, Gdańsk (1999)
313. S.L. Braunstein, C.M. Caves, R. Jozsa, N. Linden, S. Popescu, R. Schack: Phys. Rev. Lett. **83**, 1054 (1999)
314. N. Linden, S. Popescu: Phys. Rev. A **59**, 137 (1999)
315. R. Schack, C.M. Caves: J. Mod. Opt. **47**, 387 (2000)

316. E. Knill, R. Laflamme: Phys. Rev. Lett. **81**, 5672 (1998)
317. P. Slater: J. Phys. A **32**, 5261 (1999) (LANL preprint quant-ph/9806089; LANL preprint quant-ph/9810026)
318. R. Clifton, H. Halvorson: Phys. Rev. A **61**, 012108 (2000)
319. D. Deutsch, A. Ekert, R. Jozsa, C. Macchiavello, S. Popescu, A. Sanpera: Phys. Rev. Lett. **77**, 2818 (1996)
320. S.J. van Enk, J.I. Cirac, P. Zoller: Phys. Rev. Lett. **78**, 4293 (1997); Science **279**, 205 (1998)
321. T.M. Cover, J.A. Thomas: *Elements of Information Theory* (Wiley, New York 1991)
322. P. Shor: Phys. Rev. A **52**, 2439 (1995)
323. C.H. Bennett, D.P. DiVincenzo, J. Smolin: Phys. Rev. Lett. **78**, 3217 (1997)
324. D. Bruss, D.P. DiVincenzo, A. Ekert, C.A. Fuchs, C. Macchiavello, J. Smolin: Phys. Rev. A **57**, 2368 (1998)
325. M. Horodecki, R. Horodecki: Phys. Lett. A **244**, 473 (1998)
326. C.H. Bennett, H.J. Bernstein, S. Popescu, B. Schumacher: Phys. Rev. A **53**, 2046 (1996)
327. M.A. Nielsen: Phys. Rev. Lett. **83**, 436 (1999); G. Vidal: Phys. Rev. Lett. **83**, 1046 (1999); D. Jonathan, M.B. Plenio: Phys. Rev. Lett. **83**, 1455 (1999)
328. W. Dür, H.-J. Briegel, J.I. Cirac, P. Zoller: Phys. Rev. A **59**, 169 (1999)
329. D. Deutsch: Proc. R. Soc. London A **425**, 73 (1989)
330. N. Gisin: Phys. Lett. A **210**, 151 (1996)
331. B. Kraus, J.I. Cirac, S. Karnas, M. Lewenstein: Phys. Rev. A **61**, 062302 (2000) (see also M. Lewenstein, J.I. Cirac, S. Karnas: LANL preprint quant-ph/9903012)
332. M. Horodecki: presented at Dagstuhl Seminar *Quantum Algorithms* (Dagstuhl, 1998)
333. G. Vidal, J.I. Cirac: LANL preprint quant-ph/0102036
334. E. Rains: LANL preprint quant-ph/9707002
335. P. Horodecki, M. Horodecki, R. Horodecki: Acta Phys. Slovaca **48**, 141 (1998)
336. M. Horodecki, P. Horodecki, R. Horodecki: Phys. Rev. Lett. **84**, 4260 (2000)
337. K.G.H. Vollbrecht and R.F. Werner: LANL preprint quant-ph/0010095
338. D.P. DiVincenzo, P.W. Shor, J.A. Smolin, B. Terhal, A. Thapliyal: Phys. Rev. A **61**, 062312 (2000) (LANL preprint quant-ph/9910026)
339. P.W. Shor, J. Smolin, A. Thapliyal: LANL preprint quant-ph/0005117
340. P. Horodecki, M. Horodecki, R. Horodecki: Phys. Rev. Lett. **82**, 1056 (1999)
341. R. Horodecki, M. Horodecki, P. Horodecki: Phys. Lett. A **222**, 21 (1996)
342. K. Banaszek: Phys. Rev. A **62**, 024301 (2000) (LANL preprint quant-ph/0002088)
343. M. Horodecki, P. Horodecki, R. Horodecki: Phys. Rev. A **60**, 1888 (1999)
344. L. Henderson, L. Hardy, V. Vedral: Phys. Rev. A **61**, 062306 (2000) (LANL preprint quant-ph/9910028)
345. G.M. d'Ariano, H.P. Yuen: Phys. Rev. Lett. **76**, 2832 (1996)
346. D. Dieks: Phys. Lett. A **92**, 271 (1982); W.K. Wooters, W.H. Żurek: Nature **299**, 802 (1982)
347. D. Rohrlich, S. Popescu: Phys. Rev. A **56**, 3319 (1997)
348. T. Mor: LANL preprint quant-ph/9608005; T. Mor, P. Horodecki: LANL preprint quant-ph/9906039
349. N. Gisin: Phys. Lett. A **210**, 157 (1996)

350. N. Linden, S. Massar, S. Popescu: Phys. Rev. Lett. **81**, 3279 (1998); A. Kent: Phys. Rev. Lett. **81**, 2839 (1998); A. Kent, N. Linden, S. Massar: Phys. Rev. Lett. **83**, 2656 (1999)
351. D. Gottesmann: LANL preprint quant-ph/9802007
352. M. Murao, V. Vedral: Phys. Rev. Lett. **86**, 352 (2001) (LANL preprint quant-ph/0008078)
353. W. Dürr, I. Cirac: Phys. Rev. A **62**, 022302 (2000) (LANL preprint quant-ph/0002028)
354. H.-K. Lo, S. Popescu: Phys. Rev. Lett. **83**, 1459 (1999)
355. P. Horodecki, M. Horodecki, R. Horodecki: J. Mod. Opt. **47**, 347 (2000)
356. V. Vedral: Phys. Lett. A **262**, 121 (1999)
357. M. Horodecki, P. Horodecki, R. Horodecki: Phys. Rev. Lett. **84**, 2014 (2000)
358. M. Lewenstein, B. Kraus, J.I. Cirac, P. Horodecki: Phys. Rev. A **62**, 052310 (2000)
359. M. Lewenstein, B. Kraus, P. Horodecki, J.I. Cirac: LANL preprint quant-ph/0005112
360. P. Horodecki, M. Lewenstein, G. Vidal, J.I. Cirac: Phys. Rev. A **62**, 032310 (2000)
361. J. Smolin: Phys. Rev. A **63**, 032306 (2001) (LANL preprint quant-ph/0001001)
362. P.W. Shor, J. Smolin, B. Terhal: LANL preprint quant-ph/0010054
363. P. Horodecki, M. Lewenstein: Phys. Rev. A **85**, 2657 (2000)
364. R.F. Werner, M.M. Wolf: LANL preprint quant-ph/0009118
365. P. Horodecki, M. Horodecki, R. Horodecki: Phys. Rev. A **63**, 022310 (2001) (LANL preprint quant-ph/0002021)
366. B. Schumacher, M.A. Nielsen: Phys. Rev. A **54**, 2629 (1996)
367. A. Peres: Found. Phys. **29**, 589 (1999)
368. R.F. Werner, M.M. Wolf: Phys. Rev. A **61**, 062102 (2000) (LANL preprint quant-ph/9910063)
369. B. Terhal: Phys. Lett. A **271**, 319 (2000) (LANL preprint quant-ph/9911057)

Index

activation   183, 190, 193
algebra of observables   34
algorithm   29, 30, 96
– Deutsch   8
– entanglement-driven   128
– Gauss   106, 115
– Grover   30, 122
– Shor   29, 124
– Simon   113
– state preparation   10, 108
– superposition-driven   129
ancilla   9, 101, 147
ancilla form   43
antiunitary   41

basic coding principle   148
basis   154, 163, 180
– local   177
– orthonormal   156, 157
– product   156, 162, 164, 175, 176
–– unextendible (UPB)   163, 180
Bell correlation   21
Bell states   61
Bell's inequality   4, 65, 151, 155
– CHSH   4, 167
– experiments   5
Bell's Telephone   17, 19
Bell-state analysis   79
best separable approximation (BSA)   163
binary symmetric channel   142, 146
bit   152, 170
– classical   35, 170
– quantum   170
Boolean functions   8, 103
bottleneck inequality   49
broadcasting entanglement   160

capacity
– classical   47, 49, 50
– of quantum channel   194, 195
cb norm   47
channel   39, 170–172, 174
– capacity   47, 194
– classical   183, 184
– depolarizing   47, 171, 174
– ideal   48
– memoryless   171, 172
– noisy   170, 191
– quantum   146, 171, 194
–– depolarizing   171
–– one-qubit   171
– separable   42, 49
classical
– bit   35, 170
– capacity   47, 49, 50
– correlation   39
– fidelity   183
– systems   34
– teleportation   16
cloning   17
coding   170
– dense   53
– superdense   29
– theorem   34, 51
coherent information   52, 195
communication   175
– classical   152, 156, 170, 171, 188, 192
– of quantum states   152
– potential   172
– quantum   59, 62, 152, 170
– via noisy channel   170
complementarity   19
completely positive   40, 42
complexity
– classes   9, 112

– measure   99
composition   36
– of channels   40
computation path   111
continued fractions   127
controlled NOT (CNOT)   98
Cooley–Tukey decomposition   116
correlation
– classical   39
– resolver   24
counterfactual   4
– conditionals   19
criterion
– Peres   157
– PPT   156, 161, 165, 168
– reduction   165, 177, 181
– separability   155, 163, 165, 180
– – structural   193
cryptography
– quantum   27, 63, 83, 153
– – BB84 protocol   7
– – entanglement-assisted   7
– Vernam cipher   7
cyclic shift   102

decoding   170
decoherence   31, 152
decomposition
– of separable state   161, 167
– Schmidt   154, 168
delayed-choice measurements   45
dense coding   53, 66, 85
density
– matrix   35
– operator   35
depolarizing   47, 171, 174
design theory   57
digitization   31
Diophantine approximation   127
discrete cosine transform (DCT)   137
disentangling machine   161
distillation   152, 170, 171, 173, 177
– protocol   172, 173, 192
– – recurrence   188
– rate   183

eavesdropper   27
ebit   28
effect   35

eigenvalues   165
– negative   177
eigenvectors   161, 165, 177
Einstein, Podolsky, Rosen (EPR)
– element of physical reality   3
– locality   3
embedded transforms   104
energy   185
entangled   39
– coding   50
– photon pairs   74
– states   108, 120, 128, 151
entanglement   8, 23, 58, 60, 113, 151
– bound   178, 181, 183
– broadcasting   160
– distillable   178, 191–195
– distillation   164, 170, 190
– free   152, 179
– measure   161, 191–194
– nondistillable   152
– of formation   190, 191, 194
– quantifying   153, 156, 195
– splitting   161
– swapping   71, 91
– three-qubit   5
– two-qubit   178, 193
– witness   159, 160, 165, 193
entropy   156
– relative   52, 191, 192, 194
– Renýi   156
– von Neumann   174
equal distribution   115, 120, 123
error correction
– counterfactual   170
– quantum   31, 140, 171
error-correcting code
– classical   141
– CSS   149
– dual   142
– quantum   140, 147
– Reed–Müller   143

fidelity   51, 170, 171, 183–189, 192
– classical   183
filtering   175–177, 181, 186
– operation   175
– operator   175
finite field   103, 141
Fourier transform   102, 115, 130

– abelian   116
– and sampling   119
– quantum   117, 125, 129, 135
fully entangled fraction   154, 167, 173

Galois correspondence   119
gate
– graphical notation   98
– quantum   173
– quantum XOR   98, 173, 188
– Toffoli   102, 104, 105, 135
– two-qubit   161, 173
– universal set of   99
Gauss's algorithm   106, 115
Gaussian channel   31
generalized observable   42
GHZ state   5
graph
– isomorphism problem   136
– of a function   105
group
– abelian   117
– cyclic   107, 115, 118, 130, 133
– dihedral   101, 133, 136
– elementary abelian   147
– general linear   106
– quaternion   133
– solvable   130
– symmetric   100
– wreath product   133, 136
Grover's algorithm   30, 122

Hadamard
– matrices   57
– transform   9, 109, 125, 143, 147
halting problem   29
Hamming weight   141
Heisenberg picture   39
hidden-subgroup problem   127, 135
hidden-variable theory   3, 24
Holevo's coding theorem   52
Horner's rule   104
hybrid   37

impossible machines   17, 40
individual-state interpretation   23
inequalities
– Bell   4, 20, 65, 151, 155
– bottleneck   49

– CHSH–Bell   4, 167
– entropic   155, 165, 167, 181
information   152, 170–172
– classical   170
– coherent   52, 195
– mutual   195
– quantum   11, 14, 59, 170, 183
Instrument   42
interference   8, 120
irreversibility   179, 180, 194
– formation–distillation   180
– in entanglement processing   179
– preparation–measurement   180
isomorphism   159

Kraus form   43

Latin square   57
local
– realistic theory (LRT)   2
– unitary transformations   98
locality   3
loopholes
– detection   5
– locality   5

many-worlds interpretation   14, 30
map
– completely positive   157
– linear   158
– LOCC   178
– positive   157
– – decomposable   158, 167
– – nondecomposable   158, 167
– trace-preserving   158
matrix   156, 157, 175
– circulant   122, 123
– decomposing a representation   130
– encoding an error-correcting code   141
– of twiddle factors   117, 132
– permutation   100, 102, 106
– reduced density   165
maximal probability of error   50
measurement   4, 10, 175, 176
– generalized   175
mixed-state analyzer   23
mixture   40
modular exponentiation   107

214   Index

monotonicity
– condition   192
– weak   191, 192

no-cloning theorem   6, 27, 31, 184
nonlocality   151, 155, 195
– without entanglement   180
nonseparability, principle of   151
norm   175
– trace   153
norm of complete boundedness   47
normal subgroup   121, 131
nuclear magnetic resonance   168
– quantum computing   1, 169

Observable   41
observable   179
– Bell–CHSH   4, 155
– dichotomic   3, 165, 179
off-diagonal fidelity   51
one-shot
– classical capacity   50
– operation   25
operations   158
– filtering   175
– local   152, 156
– LOCC   172, 178, 190
– – entanglement-enhanced   190
– – trace-preserving   184, 185
operator   157
– filtering   175
– positive   153, 156
optimal cloning   26
oracle   8

parametric down-conversion   74
permutations   106, 116
perturbation   45
physical carrier of information   15
ping-pong ball test   20
positive   35
POVM   42
projector   153
– local   183
– two-dimensional   178
protocol   173, 176, 187–189
– BBPSSW   173, 175–177
– for distillation   172, 173, 192
– hashing   174

pure   35
purification   26, 38, 46
PVM   42

quantum
– capacity   49
– channel   146, 171, 194
– – depolarizing   47, 171, 174
– communication   59, 62, 152, 170
– computing   8, 96, 153, 168, 169
– copier   18
– correlations   152
– cryptography   27, 63, 83, 153
– – BB84 protocol   7
– – entanglement-assisted   7
– data   152
– dense coding   53, 66, 85
– entropies   156
– error correction   31, 140, 171
– error-correcting code   147
– information   11, 15, 59, 170, 183
– networks   98
– operations   173
– parallelism   9, 30
– predictions   151
– state   25, 34, 151, 152, 160, 170
– system   155
qubit   35, 49, 98, 155
qutrit   188

Radon–Nikodym theorem   45
range   161, 163, 164
rank   156, 181
– local   181
– of BE states   180
realism   3
– counterfactual   4
reduced density matrix   165
redundancy   31
relative entropy   52, 191, 192, 194
repeatability   43
representation
– induced   135
– irreducible   118, 120, 121, 130
– regular   118, 130
restriction   37, 41
ring normal form   103

Schmidt decomposition   38

Schrödinger
- cat state 108
- picture 39
Schur's lemma 120
separable 39, 47
- channel 42, 49
Shor's algorithm 29, 124
Simon's algorithm 113
singlet fraction 154
space
- Hilbert 153
- Hilbert–Schmidt 157
- real Euclidean 159
spectrum 168
spin 187
- correlations 4, 151
state 35
- bipartite 172, 183, 184, 191
- bound entangled (BE) 178, 181, 183
- classically correlated 153
- distillable 175, 177, 178, 181, 182, 185
- entangled 108, 120, 128, 151
-- maximally 154
-- nondistillable 179
- free entangled (FE) 152, 179
- isotropic 166, 173, 177, 191, 192
-- two-qubit 177
- mixed 151
-- maximally 168, 171
- NPT 156, 176, 181
- NPT bound entangled 181, 182
- PPT 156, 178
- PPT entangled 163, 168
- Werner 166, 181, 182
state estimation 25
statistical mechanics 32
statistical test 16
Stinespring theorem 44
subadditivity 191, 192
subspace 163, 164, 178
- antisymmetric 165
- complementary 180
- symmetric 165, 166
subsystem 151, 152, 155, 165, 169
superdense coding 29
superoperator 191
superposition 109, 111, 113, 125, 129

symmetry 41
syndrome 142
system
- bipartite 153, 179
- classical 34
- compound 151
- multipartite 153, 168, 179, 183
- three-level 188
- three-qubit 5
- two-qubit 161

teleportation 27, 54, 183
- classical 16
- conclusive 185–188
- experiment 89
- quantum 27, 68, 89, 170
tensor product 36
theorem
- coding 34, 51
- Hahn–Banach 159
- Holevo 52
- no-cloning 6, 27, 31, 184
- Radon–Nikodym 45
- Shannon 195
- Stinespring 44
thermodynamics of entanglement 194
time 172
time-reversal symmetry 41
Toffoli gate 102, 104, 105, 135
transpose 53
Turing machine
- configuration 110
- deterministic 110
- probabilistic 110
- quantum 111
- reversible 8
twirling 166, 173, 181, 192
two-bit gate 98, 99
two-step coding inequality 49

unitary operations
- approximations of 99
- embeddings into 104
units of information 49
universal joint measuring device 19
universal not 26

vector
- maximally entangled 154
- product 161, 163, 164

– unit   155
Vernam cipher   7
von Neumann
– entropy   51

– measurement   42

Wedderburn decomposition   130
word problem   29

Springer Tracts in Modern Physics

155 **High-Temperature-Superconductor Thin Films at Microwave Frequencies**
By M. Hein 1999. 134 figs. XIV, 395 pages

156 **Growth Processes and Surface Phase Equilibria in Molecular Beam Epitaxy**
By N.N. Ledentsov 1999. 17 figs. VIII, 84 pages

157 **Deposition of Diamond-Like Superhard Materials**
By W. Kulisch 1999. 60 figs. X, 191 pages

158 **Nonlinear Optics of Random Media**
Fractal Composites and Metal-Dielectric Films
By V.M. Shalaev 2000. 51 figs. XII, 158 pages

159 **Magnetic Dichroism in Core-Level Photoemission**
By K. Starke 2000. 64 figs. X, 136 pages

160 **Physics with Tau Leptons**
By A. Stahl 2000. 236 figs. VIII, 315 pages

161 **Semiclassical Theory of Mesoscopic Quantum Systems**
By K. Richter 2000. 50 figs. IX, 221 pages

162 **Electroweak Precision Tests at LEP**
By W. Hollik and G. Duckeck 2000. 60 figs. VIII, 161 pages

163 **Symmetries in Intermediate and High Energy Physics**
Ed. by A. Faessler, T.S. Kosmas, and G.K. Leontaris 2000. 96 figs. XVI, 316 pages

164 **Pattern Formation in Granular Materials**
By G.H. Ristow 2000. 83 figs. XIII, 161 pages

165 **Path Integral Quantization and Stochastic Quantization**
By M. Masujima 2000. 0 figs. XII, 282 pages

166 **Probing the Quantum Vacuum**
Pertubative Effective Action Approach in Quantum Electrodynamics and its Application
By W. Dittrich and H. Gies 2000. 16 figs. XI, 241 pages

167 **Photoelectric Properties and Applications of Low-Mobility Semiconductors**
By R. Könenkamp 2000. 57 figs. VIII, 100 pages

168 **Deep Inelastic Positron-Proton Scattering in the High-Momentum-Transfer Regime of HERA**
By U.F. Katz 2000. 96 figs. VIII, 237 pages

169 **Semiconductor Cavity Quantum Electrodynamics**
By Y. Yamamoto, T. Tassone, H. Cao 2000. 67 figs. VIII, 154 pages

170 **d–d Excitations in Transition-Metal Oxides**
A Spin-Polarized Electron Energy-Loss Spectroscopy (SPEELS) Study
By B. Fromme 2001. 53 figs. XII, 143 pages

171 **High-T_c Superconductors for Magnet and Energy Technology**
By B. R. Lehndorff 2001. 139 figs. XII, 209 pages

172 **Dissipative Quantum Chaos and Decoherence**
By D. Braun 2001. 22 figs. XI, 132 pages

173 **Quantum Information**
An Introduction to Basic Theoretical Concepts and Experiments
By G. Alber, T. Beth, M. Horodecki, P. Horodecki, R. Horodecki, M. Rötteler, H. Weinfurter, R. Werner, and A. Zeilinger 2001. 60 figs. XI, 216 pages

Printing: Mercedes-Druck, Berlin
Binding: Stürtz AG, Würzburg